Research and Practice in Applied Linguistics

General Editors: **Christopher N. Candlin** and **David R. Hall**, Linguistics Department, Macquarie University, Australia.

All books in this series are written by leading researchers and teachers in Applied Linguistics, with broad international experience. They are designed for the MA or PhD student in Applied Linguistics, TESOL or similar subject areas and for the language professional keen to extend their research experience.

Titles include:

Francesca Bargiela-Chiappini, Catherine Nickerson and Brigitte Planken
BUSINESS DISCOURSE

Sandra Beatriz Hale
COMMUNITY INTERPRETING

Geoff Hall
LITERATURE IN LANGUAGE EDUCATION

Richard Kiely and Pauline Rea-Dickins
PROGRAM EVALUATION IN LANGUAGE EDUCATION

Marie-Noëlle Lamy and Regine Hampel
ONLINE COMMUNICATION IN LANGUAGE LEARNING AND TEACHING

Virginia Samuda and Martin Bygate
TASKS IN SECOND LANGUAGE LEARNING

Cyril J. Weir
LANGUAGE TESTING AND VALIDATION

Tony Wright
CLASSROOM MANAGEMENT IN LANGUAGE EDUCATION

Forthcoming titles:

Dick Allwright and Judith Hanks
THE DEVELOPING LEARNER

Anne Burns
LITERACIES

David Butt and Annabelle Lukin
GRAMMAR

Alison Ferguson and Elizabeth Armstrong
COMMUNICATIONS DISORDERS

Lynn Flowerdew
CORPORA AND LANGUAGE EDUCATION

Sandra Gollin and David R. Hall
LANGUAGE FOR SPECIFIC PURPOSES

Marilyn Martin-Jones
BILINGUALISM

Martha Pennington
PRONUNCIATION

Norbert Schmitt
VOCABULARY

Helen Spencer-Oatey and Peter Franklin
INTERCULTURAL INTERACTION

Devon Woods and Emese Bukor
INSTRUCTIONAL STRATEGIES AND PROCESSES IN LANGUAGE EDUCATION

Research and Practice in Applied Linguistics
Series Standing Order ISBN 1–4039–1184–3 hardcover
Series Standing Order ISBN 1–4039–1185–1 paperback
(outside North America only)

You can receive future titles in this series as they are published by placing a standing order. Please contact your bookseller or, in case of difficulty, write to us at the address below with your name and address, the title of the series and one of the ISBNs quoted above.

Customer Services Department, Macmillan Distribution Ltd, Houndmills, Basingstoke, Hampshire RG21 6XS, England

Online Communication in Language Learning and Teaching

Marie-Noëlle Lamy

and

Regine Hampel

First published in 2007 by
PALGRAVE MACMILLAN
Houndmills, Basingstoke, Hampshire RG21 6XS and
175 Fifth Avenue, New York, N.Y. 10010
Companies and representatives throughout the world.

PALGRAVE MACMILLAN is the global academic imprint of the Palgrave Macmillan division of St. Martin's Press, LLC and of Palgrave Macmillan Ltd. Macmillan® is a registered trademark in the United States, United Kingdom and other countries. Palgrave is a registered trademark in the European Union and other countries.

ISBN-13: 978–0–230–00126–8 hardback
ISBN-10: 0–230–00126–2 hardback
ISBN-13: 978–0230–00127–5 paperback
ISBN-10: 0–230–00127–0 paperback

This book is printed on paper suitable for recycling and made from fully managed and sustained forest sources. Logging, pulping and manufacturing processes are expected to conform to the environmental regulations of the country of origin.

A catalogue record for this book is available from the British Library.

A catalog record for this book is available from the Library of Congress.

10 9 8 7 6 5 4 3 2 1
16 15 14 13 12 11 10 09 08 07

Printed and bound in Great Britain by
Antony Rowe Ltd, Chippenham and Eastbourne

To Fiona. To Stewart

Contents

Part II   Research and Practice

Part III Practitioner Research

Part IV   Resources

List of Figures

List of Tables

Checklists

General Editors' Preface

Research and Practice in Applied Linguistics is an international book series from Palgrave Macmillan which brings together leading researchers and teachers in Applied Linguistics to provide readers with the knowledge and tools they need to undertake their own practice-related research. Books in the series are designed for students and researchers in Applied Linguistics, TESOL, Language Education and related subject areas, and for language professionals keen to extend their research experience.

Every book in this innovative series is designed to be user-friendly, with clear illustrations and accessible style. The quotations and definitions of key concepts that punctuate the main text are intended to ensure that many, often competing, voices are heard. Each book presents a concise historical and conceptual overview of its chosen field, identifying many lines of enquiry and findings, but also gaps and disagreements. It provides readers with an overall framework for further examination of how research and practice inform each other, and how practitioners can develop their own problem-based research.

The focus throughout is on exploring the relationship between research and practice in Applied Linguistics. How far can research provide answers to the questions and issues that arise in practice? Can research questions that arise and are examined in very specific circumstances be informed by, and inform, the global body of research and practice? What different kinds of information can be obtained from different research methodologies? How should we make a selection between the options available, and how far are different methods compatible with each other? How can the results of research be turned into practical action?

The books in this series identify some of the key researchable areas in the field and provide workable examples of research projects, backed up by details of appropriate research tools and resources. Case studies and exemplars of research and practice are drawn on throughout the books. References to key institutions, individual research lists, journals and professional organizations provide starting points for gathering information and embarking on research. The books also include annotated lists of key works in the field for further study.

The overall objective of the series is to illustrate the message that in Applied Linguistics there can be no good professional practice that isn't based on good research, and there can be no good research that isn't informed by practice.

CHRISTOPHER N. CANDLIN and DAVID R. HALL
Macquarie University, Sydney

Acknowledgements

We would like to thank Jim Coleman for providing the initial impetus to the idea of this book and for persuading us that we could write it. We also thank François Mangenot for his incisive comments in the early stages of writing, as well as the series editors, Chris Candlin and David Hall, for their exhaustive, critical and friendly reading. We thank Fiona Doloughan for her advice on some chapters, Julie Grayson for helping us to keep track of proofs and Ruth Willats for seeing the text through to production.

Finally, thank you to Jill Lake for the email equivalent of endless comforting cups of tea throughout the process!

All errors and omissions are entirely the authors'.

Introduction

Our work as language teachers and researchers in online settings has, over the last decade, led us continually to scrutinise the field of computer-assisted language learning. However, the topic crystallised in our minds and became a plan for a book when we discovered by chance that we had each been engaged in some language learning of our own. One of us, a native speaker of German, wanted to improve her Spanish; the other, a native speaker of French, was interested in developing her Italian. We had both decided that individual learning would be the most convenient option, and we each sought a tandem partner, a Spaniard and an Italian, with an interest in improving their German and their French, respectively. But whilst one of us found her partner in the physical surroundings of the university campus and instigated tandem learning sessions in the coffee lounge, the other met her partner on a website and went on to organise sessions through Internet telephony.

When we discovered that we had had this parallel experience, we were keen to make comparisons. Did each of us feel that she had made progress in the chosen language? Yes, but self-study between sessions was a major contributor to the learning gains in both the coffee lounge and the Internet-based settings. Had the experience been pleasurable and motivating for each pair? Yes, but the atmosphere in the campus-based partnership was different from that which was created by the cyber-tandem, for whom the technology itself took the place of coffee in establishing a common ground on which to build the learning–teaching relationship. The fact that one pair used electronic technology and not the other seemed to have made little difference to the first question. However, it created two interestingly contrasting answers to the second one.

In that discussion of our experience as learners, the need to understand better the diverse ways in which technology affects learning came

1

to the fore. This need, with which our practice as teachers and researchers had already acquainted us, was once again a live issue. It is therefore from the triple perspective of teacher, researcher and learner that we offer our reflections in this book about communication in online learning, online teaching and research on online language education.

Communication is the central concept of this book. We examine theories, pedagogies and tools that in one way or another facilitate communication. Chun (2007) suggests that 'communication' used in the phrase 'computer-assisted communication' (CMC) receives the most coverage of all topic categories in her overview of recent research, based on evidence from two major US journals on technology-mediated language learning. CMC also comes top of a list of 'hits' tracked by one of the two journals in Chun's corpus. However, she adds two caveats to this apparent domination of the field. First, the acronym CMC can be used loosely to refer to a form of technology (rather than a form of communication), in which case articles categorised as 'about CMC' sometimes cover learning activities that are not communicative. Second, she notes that her sample is limited to two journals, each of which has recently devoted special issues to CMC, creating a bias in her figures.

As we show in chapter 1, such figures require further scrutiny. For example, the likelihood is that the majority of contributors to (and possibly readers of) these two journals are from the side of the digital divide where tools for CMC are becoming routine household and study items. Nevertheless, Chun is right to put the spotlight on communication, and in this book we set out to illuminate a wide range of aspects of communication in online language learning.

Part I is devoted to a discussion of the major concepts that underpin the field of computer-mediated communication for language learning. In chapter 1 we trace the origins of the field, show how it has been conceptualised and what has been expected of it over the three decades of its existence, particularly in respect of its relationship with computer-assisted language learning (CALL). In chapter 2 we identify the two major theoretical frameworks that have informed the work of our community – cognitivist second language acquisition (SLA) theory and socioculturalism – highlighting not only their influence on the technology-mediated practices in the field, but also the ways in which technology-mediated practices have been the instruments for a continuous critical reappraisal of their respective principles. In chapter 3 we come to three core concepts of education that have found a new expression and a central role in technology-enhanced learning: mediation, literacy and the affordances of technologised learning situations. Chapter 4 examines the methods used in research on computer-mediated communication for language learning, drawing out their specific advantages (or in some cases their disadvantages) for researchers in the field. In chapter 5 we turn to online pedagogies and teaching skills, showing how both types of know-how are shaped by our profession's increasing understanding of the conditions that prevail in the electronic media. Chapter 6 looks at learners and the quality of their experience, which may be facilitated or inhibited by the technological setting. In this chapter we also present ideas and questions about the role and transformations of learners' identities online currently emerging in the field. The final chapter in Part I, chapter 7, examines the assessment of online language learning in communicative settings.

Part II documents teaching, learning and research through different technologies in order to frame these examples within an overarching question about the functioning of the cycle of practice and research. By examining asynchronous fora (chapter 8), synchronous chat systems (chapter 9), multiple object-oriented environments (chapter 10), audiographic environments and virtual worlds (chapter 11), videoconferencing (chapter 12) and emerging technologies such as blogs, wikis and mobile devices (chapter 13), we bring into focus the effective or dysfunctional relationship between practice and research in our field.

In Part III we address readers interested in carrying out small-scale research on language teaching or learning in computer-mediated settings. Chapter 14 presents a picture of 'the small-scale research project' in three case studies, and offers an overview of methods and tools of relevance to such projects. In chapter 15 we turn more specifically to issues relating to researching human participants' online behaviour from the point of view of skills and of ethics. Chapter 16 addresses the methodological and practical requirements that the collection and management of electronic data place on those organising the small-scale research projects. Finally, chapter 17 suggests some practical research projects.

In Part IV, chapter 18 offers resources, mainly web-based, for readers wanting to research further the topics covered in the book.

Part I
Key Concepts and Issues

1
Historical Background

1.1 The emergence of computer-mediated communication for language learning and teaching

Computer-mediated communication (CMC) has long been of interest to teachers, learners and researchers. As early as 1989, Mason and Kaye discussed its role in different educational contexts. The title of their classic book, *Mindweave*, drew attention to the intermingling and cross-fertilisation of ideas that CMC afforded. To language professionals it soon became clear that CMC could potentially answer two needs at once: it could be the means through which teaching occurred, and it could be an end in itself. Learners could engage with the communicative aspect of their study by exchanging language online rather than in conversation classes, as they had done hitherto.

This book is about online *communicating* in the context of language learning. In this field, designations have not really stabilised, and various acronyms (see Table 1.1) have been used to cover *learning and teaching with* as well as *communicating through* computers. Some authors have striven to find differences between these acronyms, but usage has not backed them up and in practice CALL (computer-assisted language learning) and CMC have tended to dominate. To make clear our orientation to *language* learning, henceforth we use the acronym CMCL.

CMCL appeared in the mid-1990s, when institutions began to offer asynchronous text-based networking opportunities to their students. There has since been a gradual deployment of computer tools for synchronous communication, latterly including voice-based Internet telephony, across the different sectors of language education in developed countries, in distance as well as in co-located settings, justifying a symposium devoted to this form of CMCL in 2007, see SOLE symposium in Section 18.12.

Table 1.1   Acronyms in computer-assisted language learning

CALI	Computer-Assisted Language Instruction
CALL	Computer-Assisted Language Learning
CELL	Computer-Enhanced Language Learning
CBLT	Computer-Based Language Teaching
CMC	Computer-Mediated Communication
ICALL	Intelligent CALL
MALL	Mobile technology-Assisted Language Learning
NBLT	Network-Based Language Learning
TELL	Technology-Enhanced Language Learning
WELL	Web-Enhanced Language Learning

In historical accounts it is tempting to talk about trend B having emerged from trend A, or trend A having given rise to trend B. Yet the relationship between CMCL and other forms of computer-mediated communication is less tidy than this, and at different points in their history one developed as an extension of the other, but they also sometimes ran in parallel for a while, then intersected, before diverging again.

What has happened to CMCL along the way and influenced the shape that it has today? To address this question we start by examining the historical relationship between CMCL and three different strands of technology-based developments: generic educational CMC, CALL and socio-personal CMC.

With generic educational CMC, early CMCL shared an interest in the idea of learning communities, an enthusiasm for comparing online phenomena with face-to-face ones and a focus on the changing power relations among those using the medium (for example, whether some learners would dominate groups or whether shy participants would be emboldened to use their second language). In terms of research, CMCL drew from its more generic cousin a preference for research methodologies influenced by ethnography and later by discourse analysis (see chapter 4). As CMCL practitioners increased in number and their field became more visible, they were able to move on to concerns specific to language learning, such as the benefits (or otherwise) of online communicative learning for language acquisition, and for sociocultural and intercultural development. But in the course of explorations in intercultural competence, as Belz complained, many authors returned findings 'characterized primarily in *alinguistic* terms' (2003: 68; original emphasis), prompting her to advocate a move away from general socio-psychological interpretations and back to linguistic analyses of online communication. The metaphorical tracks of generic CMC and

CMCL continue to part and reconverge as the map of the domain evolves.

With CALL by contrast, CMCL has a more resolutely divergent relationship. The timeline starts in the mid-1960s, and has been influentially characterised by Warschauer and Healey (1998) as a three-phase development. Although Bax (2003) and Jung (2005) warn against taking this three-phase timeline too literally, it does provide plausible reference points, so we summarise it here in something close to Warschauer and Healey's terms. In this view, CALL gradually moves towards a so-called integrative phase, which is where what we have been calling CMCL can be located.

Concept 1.1 A view of the history of CALL

1. First comes *behaviouristic CALL*: the computer is a provider of drills to a learner who (usually) responds on an individual basis. The skills most solicited are reading and writing. Cognitive objectives predominate.
2. There is a move through the 1980s towards *communicative CALL*. The computer retains its superiority as knower of the right answer, but at that point, in parallel with the spread of interest in communicative teaching, CALL is able to use the technology for more interactive learning and greater student choice and control. The targeted skills now include speaking and listening, but machine–learner interaction is still more frequent than learner–learner computer-mediated contact, which is often limited to talk between two or three students sitting together in front of the same machine.
3. The period from 1990s through to the beginning of the twenty-first century is a phase of *integrative CALL*. Due to the arrival of multimedia products and the democratisation of Internet use, a variety of media (text, hypertext, graphics, sound, animation, video) and the written, visual or spoken productions of human beings can be accessed in an integrated fashion from a multimedia networked computer. Several skills can be deployed at once, approximating communication in non-computer-mediated environments much better. It also means that learning and teaching online can be group-based, affording the possibility that CALL can accommodate socio-cognitive and collaborative pedagogies.

CALL can thus be said to have generated CMCL as one of its constituent parts (Warschauer and Kern, 2000: 1) or termed an extension of CALL, now running on a separate track. Some proponents of CMCL reject the notion that CMCL is part of CALL, in order to escape the dominance of cognitivism, on the one hand (Harrington and Levy, 2001), and of product-driven development, perceived to be tainted with the search for economic efficacy in education to the detriment of cultural gains (Cameron, 1997: 409), on the other.

With socio-personal CMC, the relationship is only now beginning to surface. Thorne and Payne point out that

> [n]ewer tools, particularly instant messaging ... have become dominant for social and age-peer interaction. Additionally, text messaging and voice communication over cell phones abound, as does individual and group engagement with graphically and thematically sophisticated video computer games. Equivalent in importance is the emergence of ubiquitous computing; the expectation of being able to remain in perpetual contact with peers and family members either through instant messaging or cell phones. (Thorne and Payne, 2005: 379)

Godwin-Jones (2003, 2005) and Kukulska-Hulme and Shield (2004) offer surveys of relevant media, but it is left to future studies to address the question of whether CMCL educationists are embracing the technology-rich lifestyle of their learners or are missing opportunities to engage with it.

1.2   The road travelled: a broad view

Warschauer (1995) put CMCL on the map by publishing the first practitioner book on the topic. According to him, the hopes of early adopters of CMCL included giving learners the opportunity to:

- communicate with native speakers;
- communicate either one-to-one or, more innovatively, one-to-many and many-to-many;
- plan their communication;
- revisit their work, owing to the permanent traces made available to them through the technologies.

Have such expectations been met? We have interrogated the ERIC (Educational Resources Information Centre) database, which offers access to 464 journals and to a range of non-journal materials. We carried out a full-text search of the years 1992–2005 using the search terms listed on the right of Figure 1.1.

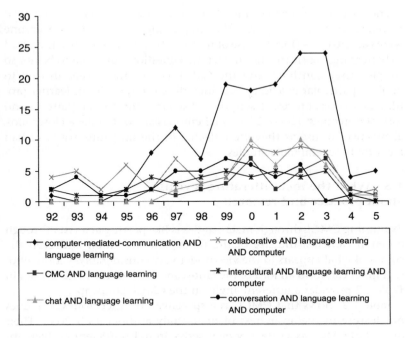

Figure 1.1 A view of the history of CMCL

Allowing for the possibility that low activity in 2004–2005 may be a reflection of the ongoing nature of database construction rather than evidence of decline, the picture shows that:

- 'computer-mediated-communication' starts to become important to language learning in 1994; and
- rises spectacularly in 1998 to reach a plateau in 2002;
- 'chat' soars from 1996;
- 'collaborative' shows a steep rise followed by a plateau, similar to 'computer-mediated-communication' in 1999;
- 'conversation' is in the ascendant from 1996; and
- 'intercultural' climbs steadily from 1996 too.

So Warschauer was right to envisage a boom in remote communication. However, to understand what this communication is for, we need to consult more detailed accounts.

The first is Warschauer and Kern's own review (2000) of the road travelled, in which they ask 'the simple question to which everyone wants an answer – Does the use of network-based language teaching lead to better language learning?' In fact, this question 'turns out to be not so simple', they conclude, and the CMCL community needs instead to 'look to particular *practices of use* including the specifics of learner profiles, task types, process description, discourse, interaction patterns and formal outcomes' (2000: 2; original emphasis). So, to take a closer look at practices of use as they are reflected in the literature, we have had recourse to several meta-studies.

1.3   What the meta-literature reveals about practice and research

Examining CMCL through meta-studies is like peering at a vista through different cameras: the size of the frame, the definition and even the colour of what appears in the viewfinder vary considerably, yet the photographer can form an idea of the landscape. Each of the meta-studies in Table 1.2 provides a different angle on the CMCL landscape.

Table 1.2 shows that CMCL as a specialism distinct from CALL does not feature in the awareness of meta-study authors until 2006. Their neglect of CMCL as a category may derive from insufficient problematisation of CMCL in the field, which may also be the reason why Jung, the author of the most extensive of our meta-studies, can assert that 'we are using the computer and the Internet as an empty transport medium like the telephone … The medium does not interfere with or impose itself on communication, it just lets through what is produced at the two ends of the line' (2005: 13). In chapters 3, 6 and 7 we hope to show just how mistaken we think his view is.

It is also clear that the diversity of characteristics of the meta-studies makes it impossible to compare them. An additional problem is that most fail to distinguish between research that reports practitioner projects in institutional settings and either experimental applications or discussions of theory. Also, the researchers do not all present their bibliographical references to show which are part of their analysis and which belong to the background, so that overlap of coverage is not easy to detect or eliminate. Finally – and this is an issue shared by all such work – practice may go unreported, particularly if results are negative. For all these reasons in section 1.4 we do not claim to provide more than trends extrapolated from the meta-information. Where the quality of research is concerned, by contrast, meta-studies are explicit, as we show in section 1.6.

Table 1.2 Meta-studies of CALL and CMCL since 1991

	Coverage	Collected	Practitioner studies only?	CMCL quantified separately from CALL?
Hassan et al., 2005	14 articles from 8 journals (plus other material)	1990–2004	Yes	Yes
Hubbard, 2005	78 articles from 4 journals	2000–2003		
Jung, 2005	5301 articles from 200 journals (plus books)	1980–2004		
Kern, 2006	36 studies from 12 journals (plus books)	1992–2005		Yes
Levy, 2000	47 articles from 5 journals (plus books)	1999 only		
Liu et al., 2002	70 articles from 21 journals	1990–2000	Yes	
Zhao, 2003	9 articles from 5 journals	1997–2001	Yes	

1.4 Practitioner studies as a reflection of practices of use

We start with Hassan et al., Liu et al. and Zhao, who explicitly declared the practice-based nature of their corpus. After narrowing down Liu et al.'s corpus to CMCL-specific material, we found that usage was overwhelmingly text-chat-based, and that Warschauer and Kern's desired 'practices of use' ('the specifics of learner profiles, task types, process description, discourse, interaction patterns and formal outcomes') were not all represented, while new ones appeared. They are listed in the left-hand column of Table 1.3 in decreasing order of frequency. The descriptors are ours.

Hassan et al. collected a wider range of CMCL literature than was discussed in their final report, which focused exclusively on real-time, voice-based studies. We subjected their entire corpus to the same treatment as above.

Table 1.3   Practice: Liu et al.'s corpus

Descriptor	Findings
Participation pattern (eight studies)	CMCL increases learner participation.
Oral skill (six studies)	CMCL promotes speaking (i.e. text chat can prepare students for speaking).
Learner experience (five studies)	Asynchronous CMCL 'received positive reactions'.
Acquisition (three studies)	Syntactic complexity in synchronous text-based CMCL: one study found less complexity and one found more.
Relational norms (two studies)	CMCL affords student-centredness and limits teacher dominance (one study); it allows native speakers to dominate in conversations involving learners (one study).
Learner attitude (two studies)	Synchronous CMCL meets with good learner attitudes.
Discourse (two studies)	CMCL allows a great variety of discourse forms.
Socio-affective skills (one study)	CMCL encourages socio-affective skills.
Sociocultural issues (one study)	CMCL involves issues of identity and community.

In both Tables 1.3 and 1.4, the first rows (where more than three studies are represented) show the same descriptors: 'participation', 'learner experience' and 'oral skill'.

We have not represented Zhao (2003)'s meta-study as a table, since it devoted only one narrative paragraph to CMCL, referencing 13 studies, most of which appear in Tables 1.3 and 1.4. Our descriptors, applied to the studies referred to in Zhao, include 'participation patterns', 'oral skill' and 'acquisition', with the addition of authenticity (two studies) and writing skill (one study). Examples of 'better planning' and 'revisiting' of work do not feature (although some do exist – e.g. Weasenforth, Biesenbach-Lucas and Meloni, 2002; Belz, 2006). This may be due as much to the difficulty of framing appropriate search strategies as to the real scarcity of such studies.

Kern (2006), an early adopter, looks back on a decade of CMCL.

Table 1.4 Practice: Hassan et al.'s corpus

Descriptor	Findings
Learner experience (16 studies)	CMCL 'received positive reactions' (11 studies). There were negative attitudes due to: CMC-based learning (three studies); technical problems (two studies).
Participation pattern (six studies)	Syntactic complexity in synchronous text-based CMCL: three studies found no difference with non-chat settings, and three found greater quantity of input but no difference in quality.
Oral skill (four studies)	CMCL promotes speaking (two studies on chat, two on audiographics).
Educational technology (two studies)	The nature of the platform influences interaction.
Literacy (two studies)	CMCL facilitates techno-literacy.
Cognitive skills (two studies)	CMCL facilitates higher-order thinking.
Teaching delivery (one study)	CMCL facilitates lesson delivery.
Discourse (one study)	CMCL facilitates communicative competence in the discourse of chat.
Task design (one study)	CMCL needs to be closely tied in to task design.

Quote 1.1 Kern's view of US-based CMCL ten years on

Research has shown that the results [of CMCL-based learning] are dependent upon a variety of social, logistical and above all pedagogical factors. In this perspective it has to be borne in mind that [CMCL] is not a genre in itself but more a collection of genres, each with its specificity, partly depending on the communication channel chosen (IRC, SMS, chatting, emailing, blogging, instant messaging, MOO) and partly due to the social and cultural context as well as the circumstances surrounding the communicative act under scrutiny ...

The pedagogical aims of teachers using [CMCL] should therefore not be limited to communication but should aim at meta-communication: exploring the relationship between language, culture, contexts and technological mediating tools.

(Kern, 2006: 27)

1.5   A new content area emerges

Kern's (2006) summary of US-based research matches in content, if not quantitatively, what we learnt in section 1.4. He charts the development of a new area – interculturalism – which is the locus of intense theory-building activity, fielding such hypotheses as:

- connectivity does not necessarily promote intercultural communication, because of the impact of institutional cultures (Belz, 2002b);
- there are cultural differences in the interaction styles of different student cohorts (Belz, 2003);
- studies must look at the impact of cultural differences on teachers in intercultural projects (Belz and Müller-Hartmann, 2003);
- researchers should investigate communicative genre and address the need to situate competence development in specific communication contexts (Kramsch and Thorne, 2001; Hanna and de Nooy, 2003; Thorne, 2003);
- success depends on interpersonal response, including mixing personal with task-related input, as demonstrated via a longitudinal study where students had a chance to become acquainted (O'Dowd, 2003, 2006);
- failure and avoidance of interaction may result when students are faced with cultural misunderstandings from a pen-pal cohort (Ware, 2003).

*

So far we have talked about the content of CMCL research. We now examine its quality.

1.6   The quality of CMCL research

Most meta-study authors (Hassan et al., Hubbard, Jung, Liu et al. and Zhao in Table 1.2) wish to see CMCL research adopt a more quantitative, experimental stance, while Levy (2000) writes at length in defence of descriptivism. Hubbard (2005) criticises researchers for reporting projects involving small numbers of untrained learners doing the task for the first time. His recommendations include better control of variables in studies through better isolation of prior experience (with the technology and the task) and finer-grained information on initial and exit proficiency (2005: 360–2). Hassan et al. use the very strict methodology of systematic reviews, which insists that studies must 'test the effect of a language learning intervention against another intervention, or standard practice or no intervention' (2005: 20). Under this criterion, not a single paper in their small corpus qualified. Liu et al. concur, observing that in their corpus 'the use of well-established measures with clear reliability and validity information was ... minimal' (2002: 263).

While a spate of research activity involving large, randomised experiments would satisfy these authors' conditions, little such research has been forthcoming. To explain this, we may postulate an *a priori* position by many CMCL researchers in favour of ethnographic work, since the object of their attention – human–human interaction – is liable to be affected by an unmanageably high number of variables. This view is supported by Bax: 'we need more careful qualitative – I would argue for ethnographic – analyses, in order to understand CALL[1] better' (2003: 2). Another argument in favour of descriptivism relates to the fact that the domain is still new. Good description lays the ground on which new theories can be erected. As Levy points out: 'Descriptive work is important in all CALL research, but especially for CMC-based work. Researchers need to be highly sensitive to the new phenomena that arise in mediated CALL learning environments' (2000: 184).

Quote 1.2 Future CALL-CMCL research should include:

... quantitative information, especially in the light of new variables emerging in recent social constructivist learning contexts, such as the role of *collaboration*, *meta-cognitive skills and knowledge* or *online presence* and *identity* ...

... qualitative and discursive syntheses of a body of research investigating similar variables related to one larger issue such as *[a given skill]* for instance, would provide comprehensive and detailed data hitherto not available.

... further high-quality, single experimental and non-experimental studies of areas relatively unexplored, such as *speaking online* ...

(Felix 2005: 286; original emphases)

The last word in this chapter comes from Felix (2005), who, while confirming the trends noted by the empiricists, suggests a more balanced range of solutions.

For a detailed exploration of some qualitative research methodologies popular with CMCL researchers, see chapter 4.

---

[1] Bax includes CMCL within CALL.

1.7 Summary

We have briefly mapped out our field, showing how some changes are interrelated and others independent of each other. We then asked what the early expectations of CMCL were and what new questions face us now that the field has been in existence for over a decade. CMCL research since the early to mid-1990s has prioritised questions on conversation and discourse, learner participation (and patterns of interaction) and collaboration (less from the point of view of task design than of learner attitude, motivation and, latterly, intercultural learning opportunities). The oral skill remains of interest, with older research looking for facilitation of oral competence in chat settings, whilst newer research observes real speech in synchronous voice-over-Internet environments. The major new content area is intercultural theory. Along with 'assessment', 'teaching delivery' continues to be under-represented.

Further reading

Chapelle (2000). In a systematic reflection on NBLT, CALL and SLA, Chapelle asks whether the arrival of NBLT (i.e. CMCL) in 2000 is an expansion or a reconceptualisation of CALL. The quality of the analysis makes it as valuable a read now as it was then.

Fotos and Browne (2004). In their introduction to this volume, which they also edited, Fotos and Browne provide a concise historical overview of CALL and CMC.

Kern, Ware and Warschauer (2004). This article provides an accessible overview of the state of the art in CMCL at the beginning of the twenty-first century, from both the teaching and the research angles.

Levy (2007). Levy reviews five perspectives on understanding cultural learning and uses different projects (email, chat, forum and Web-based) to illustrate how these perspectives can inform pedagogy.

Levy and Hubbard (2005). A brief discussion of terminologies for naming the field, and the ideological debates motivating possible choices. Levy and Hubbard come down firmly on the side of calling all CALL 'CALL', in which they include human–human mediated interaction.

Levy and Stockwell (2006). The authors examine seven dimensions of CALL: design, evaluation, computer-mediated communication, theory, research, practice and technology. The book is based on an analysis of CALL (and CMC) work constructed from the published literature between 1999 and 2005.

2
Learning Theories

Since it is difficult for most students to experience learning a second language (L2) in a natural immersion environment (i.e. where they are surrounded by the language all the time), most language teaching is done in a classroom. This instructed approach to L2 learning has given rise to the field of enquiry of second language acquisition (SLA). Two main paradigms have developed within SLA: the first is based on cognitive theories informed by psychology and linguistics; the second is influenced by sociocultural theories. Within the cognitive paradigm (which emerged first), language learning is seen as internalised – focusing on the processes within an individual's mind that can contribute to language development and on activities that help to stimulate these processes. In contrast, sociocultural theorists think of language as contextualised and see language learning as an interpersonal process situated in a social and cultural context and mediated by it. (For an in-depth discussion of mediation, see chapter 3.)

Although these theories have been developed in the context of traditional language teaching and learning in the classroom, they can also help us examine learning and teaching online. We agree with Levy that 'both theoretical positions have the potential to inform research and practice in educational computing and in CALL' (Levy, 1998: 93). Having established the centrality of these two paradigms to our field, this chapter considers how each informs concepts which have been developed around learning processes and contexts of learning – concepts which can be useful in the context of CMCL.

2.1   Theoretical framework 1: the cognitive SLA model

Cognitive SLA is an applied psycholinguistic discipline oriented towards the cognitive processes involved in the learning and use of language. It is underpinned by so-called computational models of language learning, models 'which treat acquisition as the product of processing input and output' (Ellis 2000: 194) – input being the language the learner is exposed to and output the language s/he produces. A third central concept that SLA practice and research focus on is interaction, making input, output and interaction key concepts of the cognitive approach to SLA. Figure 2.1 shows how the input–output model is structured.

Second language acquisition theory received a strong impetus from Krashen's intake or input hypothesis (1981, 1985) and his suggestion of the importance of comprehensible input for the development of a second language (see also de Bot, Lowie and Verspoor, 2005). Comprehensible input is said by Krashen to be a form of input that is just a little beyond the learner's competence but is nevertheless understood; whereas intake is 'that part of the input that the learner notices' (Schmidt, 1990: 139).

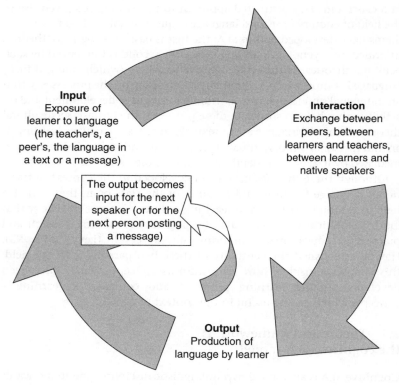

Input
Exposure of learner to language (the teacher's, a peer's, the language in a text or a message)

Interaction
Exchange between peers, between learners and teachers, between learners and native speakers

The output becomes input for the next speaker (or for the next person posting a message)

Output
Production of language by learner

Figure 2.1   Input–output model of language acquisition

Thus, for Krashen, 'the major function of the second language classroom is to provide intake for acquisition' (1981: 101) through meaningful and communicative activities. This emphasis on meaningful activities, in preference to the sole focus on form (e.g. of the grammar/translation approach), to a certain extent also takes account of the development of pragmatic competence through exposure to language in a particular context.

Other SLA researchers, however, have pointed out that Krashen's approach fails to take into account two important aspects of L2 learning: interaction and output. As a consequence, the 'interaction hypothesis' has been put forward.

Concept 2.1   The interaction hypothesis

A crucial site for language development is interaction between learners and other speakers, especially, but not only, between learners and more proficient speakers and between learners and certain types of written texts, especially elaborated ones.

(Long and Robinson, 1998: 22)

As Gass, Mackey and Pica (1998) have shown, input is most effective when it is part of interaction with others rather than with a text. Interaction allows learners to negotiate meaning, that is, to try to make meaning comprehensible (see Long, 1983; Varonis and Gass, 1985; Kramsch, 1986; Gass and Varonis, 1994). The effect of language modifications – which include simplifications, elaborations, confirmation and comprehension checks, clarification requests and recasts – is to increase input comprehensibility. These modifications 'end up providing the L2 learner with the type of negative evidence deemed necessary by some SLA theorists for continued language development' (Blake, 2000: 121). Interaction thus provides learners with the opportunity to direct their attention to language, particularly when communication has broken down.

With Swain's work, the focus in SLA broadened further to include output. Comprehensible output is seen as relevant because it provides 'the opportunity for meaningful use of one's linguistic resources' (Swain 1985: 248) and makes it possible to try out different means of expression. Swain argues that, in addition, it helps learners to concentrate on syntactic processing, that is, to focus on form. Output can trigger 'noticing', which can lead students to analyse their language and, as a result, to produce modified output. According to Swain, such monitoring contributes to acquisition.

What role can computers play in this so-called cognitive approach to language acquisition? As the use of computers has become more common in the classroom, teachers have begun to realise that CALL programmes and later CMCL applications can provide language learners not only with comprehensible input, but also with a platform for interaction where they can work with text (CALL) or negotiate meaning with peers and a tutor (CMCL). Computers also have given learners the opportunity to produce comprehensible output. Pellettieri, for example, claims that

> because synchronous NC [networked communication] fosters the negotiation of meaning and form-focused interaction … NC chatting can play a significant role in the development of grammatical competence among classroom language learners. (Pellettieri, 2000: 83)

Chapelle was one of the first researchers to call for SLA research to be used as a basis for investigating CALL (Chapelle, 1997: 28); she followed this line of investigation further in 2001. Chapelle's goals are threefold: the application of SLA theory to the design of CALL tasks; the advance of research that examines these tasks; and the development of methodological tools for that purpose. Chapelle thus suggests that the effectiveness of CALL tasks should be assessed using the following criteria.

Quote 2.1   SLA-based criteria for the effectiveness of CALL tasks

- Language learning potential: opportunity for beneficial focus on form.
- Learner fit: opportunity for engagement taking into account learner characteristics;
- Meaning focus: opportunity for focusing on meaning.
- Authenticity: correspondence of language learning activity and real-world task.
- Positive impact: effects on participants beyond language learning potential (e.g. development of language learning strategies and of literacy).
- Practicality: adequacy of resources to support the use of the CALL activity (e.g. hardware, software, technical support).

(Chapelle, 2001: 8)

Chapelle (2003) concentrates on SLA theories, showing what CALL (and CMCL to the extent that she addresses it) can offer in terms of providing input, allowing for interaction and giving the opportunity for linguistic production (or output). At the same time, she goes beyond a psycholinguistic approach to look at interaction from the perspective of sociocultural theory (which we introduce below).

Over the past decade Chapelle's call for the application of SLA theories to CALL and CMCL research has been taken up by a number of researchers. In the area of CMCL, the studies have mainly examined factors brought about by CMCL that, as Salaberry puts it, 'may contribute *indirectly* to L2 development' (2000: 6; emphasis added), concentrating on issues such as the quantity and quality of discourse. Other researchers have addressed the *direct* effect of CMCL environments on language acquisition, examining, for example, negotiation of meaning or the acquisition of grammatical structures in mediated contexts. (Examples of such studies will be introduced in Part II.)

2.2 Theoretical framework 2: sociocultural theory

Since the late 1990s there has been a general development in SLA which Block (2003) terms the 'social turn'. Influenced by the rediscovery of Vygotsky (1978) and Leontiev (1981), this approach is interdisciplinary and socially informed, and rejects 'a narrowly framed SLA whereby an overly technical model of interaction predominates ... in favour of a broader frame that integrates this narrow approach into a broader socio-culturally driven model which can account for some of the less easily defined characteristics of communication' (Block, 2003: 4). This broader frame focuses on interaction and social aspects of learning. It is important to note that in this context interaction is defined in social terms, whereas in the cognitive paradigm it is seen as 'the means by which input is made available to the black box [i.e. the human mind] or as an opportunity for producing output' (Ellis, 2003: 175).

Sociocultural theory thus offers a theoretical framework for understanding CMCL, which – with its emphasis on communication – has been claimed to provide excellent conditions for interactive, situated learning (see below). As Goodyear et al. state, 'there is no point to networked learning if you do not value learning through co-operation, collaboration, dialog, and/or participation in a community' (2004: 2). Co-construction is vital to socio-culturalism and, of course, has very significant implications for the content and the technologies of CMCL.

The growing significance of the 'social' in language learning has been fostered by developments in psychology, where, as Resnick observes, 'we seem to be in the midst of multiple efforts to merge the social and cognitive, treating them as essential aspects of one another rather than as dimly sketched background or context for a dominantly cognitive or dominantly social science' (1991: 3). For developmental processes to occur in learners' minds, social interaction is understood to be necessary.

But what exactly does 'sociocultural' mean? Sociocultural theory attempts to reconcile the analysis of psychological processes with the fact that individuals are 'situated' in social, institutional and cultural settings. Wertsch explains sociocultural theory as follows.

Concept 2.2 Sociocultural approach

The basic tenet of a sociocultural approach to mind is that human mental functioning is inherently situated in social interactional, cultural, institutional, and historical context. Such a tenet contrasts with approaches that assume, implicitly or explicitly, that it is possible to examine mental processes such as thinking or memory independently of the sociocultural setting in which individuals and groups function.

(Wertsch, 1991b: 86)

Examining the role of interaction in children's learning, Vygotsky concluded that '*human learning presupposes a specific social nature*' (Vygotsky, 1978: 88; original emphasis) since mental development is a social process before it manifests itself as an individual one. Learning therefore arises not *through* interaction but *in* interaction (Ellis, 2000). In language research, Vygotsky's concept of the 'zone of proximal development' (see Concept 2.3) has proved particularly influential and has given rise to the idea of scaffolding – assistance that is adapted to the learner's needs, generating a zone of proximal development – and to the more recent concepts of collaborative dialogue and instructional conversation (see van Lier, 1996).

Concept 2.3 Zone of proximal development

We propose that an essential feature of learning is that it creates the zone of proximal development; that is, learning awakens a variety of internal developmental processes that are able to operate only when the child is interacting with people in his environment and in cooperation with his peers. Once these processes are internalized, they become part of the child's independent developmental achievement.

(Vygotsky, 1978: 90)

Interaction with others, especially with peers and teachers, is necessary for children's learning; under adult guidance or in collaboration with more capable peers they learn to solve problems independently (Vygotsky, 1978: 86). Higher forms of learning are thus mediated, but development can only take place if mediation occurs within the zone of proximal development.

Sociocultural theory was taken up again in psychology in the late 1980s and early 1990s, and was used to critique earlier approaches to learning (e.g. Piaget's (1972) theory of cognitive development). As Lave and Wenger explain,

> conventional explanations view learning as a process by which a learner internalizes knowledge, whether 'discovered,' transmitted' from others, or 'experienced in interaction' with others. This focus on internalization ... leave[s] the nature of the learner, of the world, and of their relations unexplored. (1991: 47)

Lave and Wenger claim that, as a result, the psycholinguistic understanding of learning ignores a number of factors which also contribute to its success or failure. These have to do with learner characteristics, the role of the teacher and the setting or approach to activities (e.g. group or whole class work). Institutional and wider societal factors also play a role. We therefore need a second framework to complement the psycholinguistic model in Framework 1: namely, sociocultural theory.

Trying to understand how mental action is situated culturally, historically and institutionally, Wertsch (1991a: 15) follows this sociocultural approach. He shows how action and interaction are mediated by cultural tools such as language and how these mediational means shape the action in essential ways. But he goes further than Vygotsky and considers a second way in which human mental functioning may be socially situated: 'cognition may be viewed as being situated in broader social institutional and cultural settings' (Wertsch, 1991b: 86), such as schools and universities.

Other researchers have applied Vygotsky's ideas to second language learning. Lantolf has been particularly influential, investigating, for example, sociocultural approaches to second language learning (Lantolf and Appel, 1994; Lantolf and Thorne, 2006), the mediated mind (Lantolf and Pavlenko, 1996) and the zone of proximal development (Lantolf and Aljaafreh, 1995). Understanding second language learning as a mediated process, he examines the following domains of mediation (Lantolf, 2000):

- Social mediation: mediation by others in social interaction, e.g. mediation through experts or peers.

- Self-mediation: mediation by the self through private speech.
- Artefact mediation: by language, but also by portfolios, tasks and technology.

Although Vygotsky investigated children's learning and development, his idea that social structures influence socio-cognitive activities has informed more general research into learning, contributing to the development of a number of concepts: situated learning, ecological perspectives on language socialisation and communities of practice.

Lave describes situated practice as follows.

Concept 2.4   Situated social practice

Learning is seen as situated in 'social practice in the lived-in world ...

This theoretical view emphasizes the relational interdependency of agent and world, activity, meaning, cognition, learning, and knowing. It emphasizes the inherently socially negotiated quality of meaning and the interested, concerned character of the thought and action of persons engaged in activity. ... This view also claims that learning, thinking, and knowing are relations among people engaged in activity *in, with, and arising from the socially and culturally structured world*. This world is itself socially constituted.'

(Lave 1991: 67)

Knowledge is thus constructed in joint activity, and learning is a process of participating in cultural and social practices. This process structures and shapes cognitive activity. Learning a foreign language therefore means engaging with the L2 environment, an environment that is constituted by cultural, societal and institutional practices. Van Lier (2002) and Kramsch (2002), for example, use the metaphor of 'ecology' to describe 'the poststructuralist realization that learning is a nonlinear, relational human activity, co-constructed between humans and their environment, contingent upon their position in space and history, and a site of struggle for the control of social power and cultural memory' (Kramsch, 2002: 5). Van Lier explains that during communicative acts,

> [w]ords are used, but these words function only in conjunction with gestures (a pointing finger), gaze and the parts of the physical surroundings staked out. The whole scene can be referred to as semi-otic action, and in this semiotic action language emerges and becomes a constitutive part. So, speaking is always a part of a context

of meaning-producing actions, interlocutors, objects, and relations among all these. In other words, language emerges as an embodied and situated activity. (2002: 146)

We explore multimodality and meaning-making further in chapter 3.

Whilst all language learning communication is situated, there has been a particular interest among some CMCL practitioners in ensuring that the educational process makes the best possible use of the situation in which the learning takes place. Erben (1999), for example, used and evaluated audiographic technology to train language teachers to teach Japanese. His student-teachers 'not only engaged in reconstructed pedagogical and linguistic behaviour unique to the audiographic environment but also clearly became more self-regulating in doing so' (1999: 243). Both the behaviours that he describes – adaptive reconstruction and self-regulation – could be said to evidence situated learning, the former by orienting to what the tool uniquely had to offer, the latter through strategies of conscious autonomisation, strategies that are particularly well suited to survival in a language-immersion setting such as the one that Erben's networking project created (see also chapter 12).

The construct of communities of practice originates in the concept of situated learning. Lave and Wenger argue that 'rather than learning by replicating the performances of others or by acquiring knowledge transmitted in instruction ... learning occurs through centripetal participation in the learning curriculum of the ambient community' (1991: 100) They call this a 'community of practice'.

Concept 2.5   Community of practice

A community of practice is a set of relations among persons, activity, and world, over time and in relation with other tangential and overlapping communities of practice. A community of practice is an intrinsic condition for the existence of knowledge, not least because it provides the interpretive support necessary for making sense of its heritage. Thus, participation in the cultural practice in which any knowledge exists is an epistemological principle of learning.

(Lave and Wenger, 1991: 98)

A community of practice can give identity to its participants, i.e. a group who share 'a concern or a passion for something they do and learn how to do it better as they interact regularly' (Wenger, 2005: 1). The metaphor 'community of practice' has been useful to many CMCL researchers as a way of conceptualising the learning of online groups

(Doering and Beach, 2002; Godwin-Jones, 2003; Morita, 2004; Lewis, 2006). It should be noted that critiques of the construct have also emerged from both outside educational CMC (Gee 2000; Candlin and Candlin, 2007) and from within (Lea, 2005), but that CMCL has yet to develop it own critical position relating to this issue.

In sum, from a sociocultural perspective, CMCL can be useful for providing learners with the opportunity to interact and collaborate with others in many ways. Building collaborative language learning communities via computer networks is seen as a way of developing not only learners' linguistic skills but also cultural understanding and critical awareness (see section 1.5). Computer conferencing can connect learners, teachers and other competent speakers of the L2 and offer new possibilities for practising the language and learning about other cultures and backgrounds. It has the potential to encompass different pedagogical principles – situated practice, overt instruction, critical framing and the production and transformation of knowledge (Gee, 2000). However, within CMCL critical voices have also been heard. In response to the field's enthusiasm for connecting diverse groups of learners via the Internet and the assumption that such exchanges bring about intercultural benefits, Belz (2002a), Belz and Thorne (2006) and O'Dowd (2006b) have been calling for stronger conceptualisation of terms such as 'culture' and 'intercultural'.

Warschauer opened up CMCL research to include these sociocultural perspectives. In an influential article, he describes one of the main purposes of his work as being 'to explore the nature of computer-mediated communication (CMC) by using a conceptual framework that starts with well-known theories of input and output and leads to sociocultural learning theory' (1997: 470). Like Kern (2000) he encompasses both a cognitive and a sociocultural approach. From a cognitive perspective, the place of CALL and CMCL is to provide language input and analytic and inferential tasks; from a sociocultural perspective, their place is to provide contexts for social interaction; to facilitate access to existing discourse communities and create new ones.

Yet what are the implications of such ideas in practice? It is a very different process collaborating at a distance rather than face-to-face; creating a community with people one has never met other than virtually, and constructing one's identity in front of a screen. Do all learners feel they can contribute online? Are learners really in a position to produce and transform knowledge? What about issues such as motivation and anxiety?

Finally, with the opening up of the field to include sociocultural learning theories, there has also been an impact on the methodologies used in research. Debski and Levy point to the necessary expansion of research agendas.

> Our desire to understand what is happening when a learner or a group of learners are using a computer has been extended to involve situations where learners collaborate over distance and interact with virtual communities to accomplish creative goals. Research agendas are expanding to include issues of social computing and networked cultures and specific methodologies such as ethnography and ethnomethodology, designed to further our knowledge in this area. (Debski and Levy, 1999: 8)

2.3 Summary

We have looked at the two main influences on language learning and teaching research and practice. One has its focus on the processes of acquisition of language as they affect individuals (cognitive framework); the other centres on the learning that interacting individuals create through their very interaction (sociocultural framework). We have shown how both, but particularly the latter, provide principles to guide practice and research in our field. Sociocultural theory has also shaped CMCL practitioners' thinking about issues to which we will return: tutor roles, task design and collaborative learning (chapter 5) as well as student participation in online environments, learner motivation and anxiety, and questions of identity (chapter 6).

Further reading

From CMCL

Chapelle (2001). This important book offers a full description of the relationship between SLA theory and computer-assisted learning. However, it focuses solely on cognitive SLA and ignores what socio-constructivist theories could contribute.

Chapelle (2003). Although the focus of this book is English language learning and the use of CALL (which, for this author, includes CMC), it is also useful for readers interested in technology and language learning more generally. It examines questions around technology in the context of language teaching practices, SLA and applied linguistics.

Doughty and Long (2003). This article examines the use of technology in foreign language teaching from a psycholinguistic perspective. The authors introduce ten methodological principles of task-based language teaching and illustrate each of them with a variety of pedagogic procedures.

Lund (2006). Lund exemplifies how a researcher can make use of sociocultural theory in order to conceptualise 'communicative opportunities' in an online exchange between EFL learners and their teacher, and to identify possible enhancements to the understanding that teachers may develop of their learners' multiple online cultures.

From other fields

Block (2003). By examining the key components of SLA (second, language and acquisition), Block suggests a more interdisciplinary and socially informed approach to SLA research.

de Bot, Lowie and Verspoor (2005). A comprehensive introduction to SLA research, this book is organised in three sections and includes a series of tasks. Section A (Introduction) explains key terms and concepts, section B (Extension) presents a selection of excerpts from important publications, and section C suggests small-scale investigations.

Smith (2005). This rigorous article offers an example of cognitivist research. Smith provides both quantitative and qualitative analyses of CMCL data, set against the theoretical background of uptake theory, leading to a clear articulation of the ways in which the networked medium affects the learning.

Tella and Mononen-Aaltonen (1998). This article explains from the point of view of cognitive psychology why interaction has become so important in online learning.

Zuengler and Miller (2006). Zuengler and Miller chart the debate between cognitive and sociocultural understandings of language learning over the past 15 years.

3
Mediation, Multimodality and Multiliteracies

In this chapter we start by exploring the sociocultural concept of mediation. Human learning is mediated through interaction with others, using language as well as other 'mediational tools'. In our context these are 'participant interaction', 'tasks' and 'technology'. While the new technologies have been developed to offer modes that resemble those used in face-to-face environments (speech, writing, image, etc.), the computer medium allows for different affordances. We define this concept and examine the modes and affordances of different CMC environments and discuss the implications of these affordances and differences for language learning. The focus of the last section is the change from 'page to screen' (Snyder, 1998) which has resulted in the development of the concept of multiliteracies.

3.1 What is mediation?

To clarify the meaning of the term 'mediation', we briefly look at its origins and connection with the related term 'media'. Etymologically, 'mediation' and 'media' both refer to being 'in the middle', from the Latin *mediare* (English 'stand in the middle') and *medium* (English 'middle'), respectively. The 'media' are 'in the middle' since they are the means of getting a message from producer to receiver(s). Both 'medium' and 'mediation' have changed their meaning over time, and there is now a variety of ways to understand these terms. While 'media' is used primarily in a transmissive sense today to designate the means of mass dissemination of messages and their content (e.g. the 'television medium'), it also has a semiotic definition as a means of expression (as in oil for painting) (Ryan, 2003).

To explore the many meanings of 'mediation', we input the term into a clustering search engine. This returned a list of conflictual processes – dispute resolution, conflict, divorce, family, labour relations. From this test it is clear that mediation is widely understood as referring to interaction and involving negotiation, and that its educational meaning is far less well known. The construct of 'mediation' in education has its roots in the sociocultural theory of learning, and particularly in the work of Vygotsky (1978), Leontiev (1981) and Wertsch (1991a) (see also chapter 2). In contrast to the cognitive model of learning, sociocultural approaches stress the central role of social interaction for learning: all human learning is mediated through, or shaped by, interaction with others, and this shaping does not takes place in a vacuum but through *mediational tools*. These include:

- the language that humans use (e.g. Spanish, sign language, musical notation, Morse code);
- the cultural assumptions that they bring to the event (their belief system);
- the social institutions within which the event is taking place (e.g. a school, park, market, home);
- the software or hardware humans have at their disposal (e.g. the Internet, newspaper, abacus);
- the time structure that frames their encounter (continuous in a real-time frame, interrupted in a time-delayed one).

The 'shaping' that takes place through these mediational tools is cyclical: they help to create the learning and in turn the learner may shape these tools or exploit them for his or her own purposes. For example, early blog writers were diarists, but the creativity of bloggers has been such that a range of genres now exists, including research blogs, blending personal, social and scientific interests (see the Humlab blog at Umeå University: http://blog.umlab.umu.se/). By 'mediation', in this book we refer to this mutual shaping. Like Wertsch (2002), we therefore see this shaping as transformative.

Quote 3.1   Wertsch's view on action and transformation

[With] the introduction of a new cultural tool into the flow of human action we should be on the lookout for qualitative transformation of that action rather than a mere increment in efficiency or some other quantitative change.

(Wertsch, 2002: 106)

Language mediates all relationships

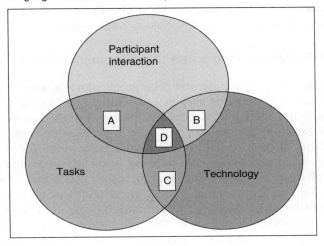

Figure 3.1   A model of mediation in CMCL

Three aspects of this transformative type of mediation inform our book, identified in Mercer, Littleton and Wegerif (2004: 203) as central to CMC-based learning. The first is the way language learning is mediated for the online learner by 'ideas from other participants'. The other two are the way CMC technology and tasks mediate learning. Figure 3.1 is inspired by Mercer et al.'s conceptualisation and represents the overlapping areas in which mediation operates during networked learning.

Language is the main mediational tool in all social human learning, particularly in language learning, where it constitutes the end as well as the means. In Figure 3.1 the centrality of language is represented by shading. The four areas A–D represent the meeting points of the three mediational tools: 'participant interaction', 'tasks' and 'technology'. For example, A represents the mutual impact of interaction-type and task-type, giving rise to possible research questions such as: Given a form of participant interaction (e.g. collaboration) and a task type (e.g. problem-solving), how does the conjunction of the two mediational means shape the learning, and will the quality of that learning change if we vary the means (e.g. substitute problem-solving with presentation or with role-play)? B represents the junction between forms of participant interaction and particular technologies. In C the focus is on the mutual shaping between particular task-types and specific technologies. In this book, we are particularly concerned with D, where the technological means are a consideration underpinning the construction of all research questions, i.e. the influence of the 'how' (means/medium) on the 'what'? (product, event, outcome).

3.2   Affordances, modes and the computer medium

The medium of the computer enables users to employ a range of ways of communicating, including spoken and written language, images and gestures. While these modes seem reminiscent of those used in the traditional face-to-face classroom (e.g. text in an exercise book or on a whiteboard; images in a book or video; spoken language face-to-face or on a tape), the new media have different possibilities and limitations. The implication of Wertsch's observation (Quote 3.1) is that conditions are transformed, so that teachers and learners cannot simply replicate in CMCL what they have become accustomed to doing in the face-to-face setting (although, we might add, they often expect to be able to). In these transformed conditions, Svensson's (2004) concern is critical: do designers work in such a way that 'traditional classrooms are often virtualized, with their "old" structures'?

The next sub-section explores the notions of affordances and modes in general terms before investigating the practicalities of using multimodal environments in the context of language learning.

3.2.1   Affordances

Affordance is a concept of wide application, originating in psychology and specifically in Gibson's (1979) research on visual perception.

Concept 3.1   Affordance

The affordances of the environment are what it offers the animal, what it provides or furnishes, either for good or ill. The verb to afford is found in the dictionary, but the noun affordance is not. I have made it up. I mean by it something that refers to both the environment and the animal in a way that no existing term does. It implies the complementarity of the animal and the environment.

(Gibson, 1979: 127)

The emphasis in Gibson is on what the animal – or the human being – perceives rather than what is inherent in the object. So although bathers might perceive a lake as somewhere to swim, it affords a fisherman a different use, such as earning a living. Affordances thus make up the different possibilities and constraints in the environment, which give agents different options for action. This environment includes not only physical objects but also social phenomena such as interaction or tools like language. An artefact can be designed to afford interaction in a learning context, for example. However, if there is a clash between this artefact and the task, these affordances are unlikely to lead to interaction (Finneran and Zhang, 2003).

Because of its focus on the relationship between environmental properties and humans, the concept of affordance has also influenced recent developments of an ecological view of learning. Based on the view that 'an affordance is a property of neither the actor nor of an object: it is a relationship between the two', van Lier (2000: 252) prioritises relational and interactional processes over material objects and products. This leads him to a critique of the psycholinguistic, input–output approaches to language learning. He suggests the *ecological* notion of affordance as an alternative to the concept of input and emphasises the idea that the unit of analysis in research should not be 'the perceived object or linguistic *input*, but the active learning, or the *activity* itself' (van Lier, 2000: 253). To understand the activity, one must, of course, understand the ecology. This includes motives and goals for action – an area that activity theory explores.

> Motives can be biologically determined, for example, the need to satisfy hunger, or ... socially constructed, for example, the need to learn an L2. The learners' motives determine how they construe a given situation. Thus people with different motives will perform the same task in different ways. (Ellis, 2003: 183)

Laurillard et al. (2000) applied the notion of affordance to research on computer-mediated learning by studying students' relation to a CD-ROM. Their study shows how, instead of discussing the content of the CD-ROM (that is, the input), the students' interaction was limited to talk about how to find their way round the software. The designers had transferred the activities from the face-to-face classroom to the electronic setting without taking into account the affordances of the media, which included navigational aspects, poorly understood by the learners because unsupported through the CD-ROM's activities. Laurillard et al. identify affordances of CD-ROMs that potentially support the learning process, such as the possibility of repeating activities, giving feedback and offering tools for reflection. Only on the basis of this can multimedia activities be designed that potentially stimulate the use of these affordances. So, as the researchers show, rather than succumbing to the limitations of the computer environment (such as the inability of the programme to recognise the students' lack of engagement with the material), a CD-ROM activity could harness the computer's potential, such as the availability of an editable Notepad.

Human learning is 'a process of making meaning – a "semiotic" process; and the prototypical form of human semiotic is language' (Halliday, 1993: 93). So language is 'not a "domain" of human knowledge ... language is the essential condition of knowing, the process by which experience "becomes" knowledge' (1993: 94). If electronic resources are to be used successfully for learning, there has to be an understanding of their 'meaning-potentials' (Kress, 2003), that is, an attention 'to the *materiality* of the resources, the material *stuff* that we use for making meaning' (2003: 32; original emphasis). In CMCL it is the 'material stuff' of the computer (hardware and software), as well as the total environment (the location in which the learner operates), that teachers and researchers need to attend to, as much as to the human aspects of the learning experience. To ensure that CMCL provides more than a pale imitation of face-to-face interaction, it is crucial that software and materials designers are aware of the ways that affordance works. If designers attend to the 'communicative affordances' of technologies, as Hutchby (2001) calls them, that is, if they design the tools for communication (text, image or icon) to best effect by taking into account how learners make use of them, it is more likely that learners will be encouraged to interact and communicate with their teachers and with one another.

We now explore the technical functionalities that the new online media offer. They include:

- browsing;
- artefact creation;
- artefact manipulation;
- displaying/storing/retrieving artefacts;
- shared textual/visual/graphic tools;
- clickable icons for interaction;
- asynchronous sending/receiving (e.g. forum);
- synchronous sending/receiving (e.g. chat);
- voice-over Internet (e.g. audio conferencing);
- simultaneous use of different channels (e.g. audiographic/video-conferencing).

These functionalities and material characteristics have affordances which in turn impact on interaction. The new media may offer seemingly familiar resources for meaning-making (written text, sound, images), but communication is mediated by the computer. This means two things: all resources have to be accessed via the computer medium (with the help of hardware such as keyboard, mouse or screen, and of software); and they are delineated by its possibilities (by features such as frames, hot buttons, pop-up menus, voting buttons, clickable options, push-to-talk devices).

3.2.2   Modes

The functionalities of the computer medium also have an impact on the modes of communication. In order to clarify our use of the term 'mode', we draw on social semiotics, examining modes in the context of meaning-making as suggested by Kress and others (e.g. Kress and van Leeuwen, 2001; Kress et al., 2001; Kress, 2003).

Concept 3.2   Modes

We define modes as 'semiotic resources which allow the simultaneous realisation of discourses and types of (inter)action' (Kress and van Leeuwen, 2001: 21). Further, each mode 'involve[s] a system with an underlying grammar (in the widest sense of this term) that embodies the organization of the system and the uses to which it can be put'.

(Chanier and Vetter, 2006: 3)

Language is not the only semiotic mode at our disposal. Representational systems go beyond written and spoken language to include images; and even written text includes visual aspects (e.g. layout, font, colour). Thus we communicate using a complex system made up of written, spoken and visual resources, each with its own modes and affordances.

In the case of CMC, the earliest tools such as email or fora offered mainly one mode: written language. Now computers can provide access to environments bringing together a number of modes, including those based on text, speech, gestures, images and icons. Even the spatial organisation of the screen plays a role in how meaning is made, so in that sense the choices made by designers in how the screen space is used also constitute a mode. Combining these in multimodal software allows for an 'orchestration of meaning' (Kress et al., 2001: 25).

Two contrasting positions have emerged: some researchers point to the potential quantitative increase in communication that such environments allow, while others perceive computer environments as restricted compared to face-to-face settings. Cook (2003) argues for the latter view:

> What has been gained through new technology is an enhanced ability to relay acts of communication; what has been lost are the modalities[2] of objects, bodily presence, timing, space, weight, temperature, light and dark, touch, taste, smell, inebriation, and internal bodily sensation. What remains are often reduced acts in which the only modalities are those of writing, vision, and sometimes sound. 'Bi-modal' or 'tri-modal' might be more accurate terms for them than 'multimodal'. (Cook, 2003: np)

While recent technologies increasingly incorporate sound, most virtual environments are indeed disembodied environments, unable to replicate modes such as gestures or touch. Humans are accustomed to using body language to accompany the mode of spoken language and make meaning in face-to-face interaction. Superficially, therefore, the new media might seem to offer nothing more than 'reduced acts' (Cook, 2003).

However, we argue that neither view is helpful. Both approach online communication on the basis of face-to-face interaction rather than trying to understand what the new media can offer in their own right. Wertsch (2002) criticises those who suggest that computer mediation provides only a quantitative increase in efficacy.

Quote 3.2   Transformative impact of a new tool

Rather than viewing the introduction of a new cultural tool as making an existing form of action easier or more efficient, it may be important to consider how it introduces fundamental change – sometimes to such a degree that we can question whether the same form of action is involved at all.

(Wertsch, 2002: 106)

This quote could also easily be adapted to counter the second view. So, rather than starting with the old form of action (e.g. the tools have no way of supporting learners shaking hands), we can understand Wertsch's comment as an encouragement to critics to consider the new form of action (e.g. the tool offers simultaneous one-to-many speaking and writing) and any of the affordances the tool may have.

---

[2] Cook uses the word 'modality' where we have been using 'mode'.

As early as 1990, Harasim recognised that CMC facilitates collaboration while acknowledging that it creates 'a new environment for learning' in which 'educational interaction is mediated' (1990: 42). Today, computers offer multimodal communication and networking tools which can encourage co-operation and collaboration in various modes, thus supporting sociocultural approaches to learning. So, by allowing learners and tutors to communicate virtually via written and spoken language, images and/or gestures, the new technologies can be seen as contributing to a fundamental change in representation and communication, as noted by Kress (2003).

Quote 3.3 Extent of impact of a new tool

[New technologies] change, through their affordances, the potentials for representational and communicational action by their users; this is the notion of 'interactivity' which figures so prominently in discussions of the new media.

(Kress, 2003: 5)

In section 3.2.3 we reflect on the implications of the notion of affordance for language learning. This leads us to a consideration of the development of a new type of literacy, which we discuss in section 3.3.

3.2.3 Online tools and affordances in CMCL

In this section we explore the impact of the functionalities, modes and affordances of CMCL. Even though most of the computer's material functions (e.g. object manipulation or sound transmission) have been developed for generic use, there are many ways of using them for CMCL. We exemplify these in Table 3.1, which examines the benefits as well as the challenges offered, while in Table 3.2 we present distinct online tools and their modes and affordances for CMCL in general and online communication in particular.

One obvious feature of CMCL is that all tools have to be accessed via the computer. This means knowing how to use the hardware (e.g. keyboard, screen, mouse) as well as the software. Table 3.1 shows other material features of the online setting and points to some of the implications for teaching and learning, taking into account both what the computer facilitates teachers and learners in doing and what the challenges are.

Table 3.1 Features of CMC environments and their uses and pitfalls for CMCL.

Material functions of online setting	In CMCL, these functions have been used in order to:	In CMCL, these functions can pose the following challenges:
Browsing	Allow access to authentic material in L2	Verifying the reliability of content can be a problem; linguistic difficulties
Artefact creation	Produce texts, images, etc.	Fairly high level of technoliteracy required
Artefact manipulation	Work with texts, images, etc.	Fairly high level of techno-literacy required
Displaying/storing/retrieving artefacts	Self-reflect on own learning	Profusion of material; threading
Shared textual/visual/graphic tools	Produce texts, images, etc. jointly	Fairly high level of techno-literacy required
Clickable icons for interaction	Foster communication and interaction (e.g. vote button); create tele-presence; help with turn-taking (e.g. raised-hand button)	Icons can be used for purposes they were originally not intended for (creatively/abusively)
Asynchronous sending/receiving (forum)	Foster peer collaboration; give feedback; meta-commenting (extended, reflective commentary)	Netiquette skills required; unequal learner participation patterns; control of volume of postings; threading structure needs to be adequate to content structure
Synchronous sending/receiving (e.g. chat)	Foster peer collaboration; give feedback	Chat literacy required; unsuitable for in-depth discussions; unequal learner participation patterns
Voice-over Internet (e.g. audio conferencing)	Foster oral communication and interaction; work collaboratively; give feedback	'Netiquette' required (e.g. for turn-taking); loss of flexibility in time; unequal learner participation patterns
Simultaneous use of different channels (e.g. audiographic/videoconferencing)	Different tools and modes can complement one another	Fairly high level of techno-literacy required; cognitive overload

Table 3.2 Modes and affordances of different CMC environments and their uses for CMCL

Online tools	Structure of time (Asynchronous = Asy. or Synchronous = Sy.) → affordance (Aff.) linked to time	Modes (M.) → affordances (Aff.) linked to mode
Fora	Asy. → Aff. 1: allows time for reflection on language. Opportunity to digest messages or to formulate responses, using other tools such as dictionaries. Can help learners who suffer from anxiety when speaking.	M.: linguistic (writing) → Aff. 4: allows focus on form to foster accuracy. Can foster analytic skills (e.g. structuring). Can foster metacognitive skills (e.g. self-reflection and revision). M.: spatial-visual (threading, nesting, etc.) → Aff. 5: allows organisation of contributions according to themes and/or time
Blogs	Asy. → Aff. 1	M.: linguistic (writing) → Aff. 4 M.: spatial-visual (structural functions within blog, e.g. sequencing, hyperlinking) → Aff. 5 M.: visual (images input by user) → Aff. 6: complement and illustrate verbal modes Can act as stimuli for interaction.
Chat	Sy. → Aff. 2: allows real-time exchanges. Rehearsing speaking	M.: linguistic (writing) → Aff. 7: if using Save Chat function, can foster metacognitive skills (e.g. self-reflection and revision). M.: linguistic (icons) → Aff. 8: socio-affective interaction M.: spatial-visual (structural functions within chat, i.e. sequencing, identifiers) → Aff. 5
MOOs (Multi-user domain, object-oriented)	Sy. → Aff. 2	M.: linguistic (writing) → Aff. 7 M.: spatial-visual (structural functions within environment, i.e. buildings metaphor, identifiers for fictive identities) → Aff. 9: supports socio-affective interaction
Audio conferencing	Sy. → Aff. 3: permits naturalistic oral conversations	M.: linguistic (speaking) → Aff. 10: using oral skill including intonation, stress pattern and expressive vocalisation.

Continued

Table 3.2   Continued

Online tools	Structure of time (Asynchronous = Asy. or Synchronous = Sy.) → affordance (Aff.) linked to time	Modes (M.) → affordances (Aff.) linked to mode
Audiographic conferencing; virtual worlds	Sy. (or quasi-sy.) → Aff. 3	M.: **linguistic** (speaking) → **Aff. 10** M.: **linguistic** (writing in chat) → **Aff. 7** M.: **linguistic** (writing on shared documents, whiteboards) → **Aff. 11**: collaborative skills M.: **spatial-visual** (structural functions within environment, i.e. rooms, games or avatar metaphors) → **Aff. 9** M.: **visual** (images input by user) → **Aff. 6**
Video-conferencing	Sy. (or quasi-sy.) → Aff. 3	M.: **gestural** (facial expressions and gestures) → **Aff. 12**: gestural output supports communication M.: **linguistic** (speaking) → **Aff. 9**

So far, we have only looked at generic features of the online setting, but what about specific online environments and the modes and affordances they offer? There is a whole range of tools, from asynchronous written conferencing (which is dominated by one mode) to videoconferencing (which affords a number of verbal and non-verbal modes of communication). Table 3.2 shows the most widely used online tools and examines their affordances in relation to both their time structure and the modes they offer.

All these environments, and particularly the first four, have been described as 'reduced-cue' (depriving users of some of the affordances of face-to-face interaction such as body language) and much research as well as pedagogical effort has been expended in attempting to account and compensate for this perceived impoverishment. However, some have argued that even in written environments, interaction can 'surpass ... the level of affection and emotion of parallel (face-to-face) interaction' (Walther, 1996: 30). This has been called 'hyperpersonal' communication (Jones, 2004).

3.3 New literacies

3.3.1 Definition of multiliteracies

Concerning affordances, user perceptions are more pertinent than the features of the object itself (see section 3.2.1). So it is not just the material affordances of CMC that play a role in enhancing or limiting communication, but also how people see them and the practices that result from their different perspectives. The notion of literacy has served to conceptualise this understanding (by users) of the tools in their environments. The OECD (2000) definition invites us to look on literacy as 'the ability to understand and employ printed information in daily activities, at home, at work and in the community – to achieve one's goals, and to develop one's knowledge and potential'. Therefore, new literacies can be seen as the knowledge and understanding that users bring to the activities they carry out in the new electronic media, empowering those on the right side of the digital divide to choose the appropriate language to represent their meaning (see also Lessig, 2004). As Kress and van Leeuwen (2001) have established in reference to non-electronic media, literacy is not limited to decoding written language; it includes understanding and using images as well as sound. We suggest that these various modes are also central to the new literacies.

The impact of technological developments on literacy in recent years can thus be summarised as a 'revolution in the uses and effects of literacy and of associated means for representing and communicating at every level and in every domain' (Kress, 2003: 1). As a result, Kress and the New London Group have called for the development of 'multiliteracies'.

Concept 3.3 Multiliteracies

Literacy pedagogy now must account for the burgeoning variety of text forms associated with information and multimedia technologies. This includes understanding and competent control of representational forms that are becoming increasingly significant in the overall communications environment, such as visual images and their relationship to the written word.

(New London Group, 1996: 60)

Other researchers talk about 'electronic literacy' (Warschauer, 1999a), 'techno-literacy' (Erben, 1999), 'technological literacies' (Lankshear et al., 1997) – though none of these labels reduces literacy to the technological aspect – or 'new literacies' (Lankshear and Knobel, 2003a). The different layers of meaning we have just explored are best encompassed by the term 'multiliteracies', which will be used in this book in the context of CMCL.

Having access to means of representation potentially has a transforming and democratising effect, empowering users and turning them into agents. 'In a semiotic(-linguistic) theory of transformation and remaking[,] the action of the individual is that of the changing of the resources ... [in] the maker's interest' (Kress, 2000: 156). An illustration of this can be found in Palfreyman and al Khalil (2003), who describe Arab students using the Latin alphabet in class to write vernacular Arabic for 'secret' online messages which the teacher cannot understand. Because today's digital networked technologies give users easy access not only to the means of representation and production but also to worldwide dissemination, this effect can be even greater.

Yet the concept of multiliteracies also has a critical dimension – what the New London Group (1996) calls 'critical framing'. The OECD's definition of literacy mentions only 'understanding and employing' information, but literacy also needs to include the awareness that representational resources are social practices constructed by a particular society and are therefore limited (Lankshear and Knobel, 2003a). Canagarajah (2002) points out that the afore-mentioned democratising effects are not an automatic adjunct to the new media.

Concept 3.4   Critical literacy

What is important is to make them [students] aware of the implications and ramifications of new literacies. The liberating and democratizing powers of current technology should not be taken for granted. Students have to go beyond developing a simple functional literacy in the new media and new genres. They have to still adopt a critical literacy for expanding the possibilities of the new resources, appropriating the available media for their oppositional purposes, and democratizing the cyberworld for broader participation.

(Canagarajah, 2002: 222–3)

3.3.2   Multiliteracies: practical challenges
for CMCL users

In the context of online communication, multiliteracies include the skill of using the hardware and software. They also involve an awareness of and ability to deal with the constraints and possibilities of the medium. Luke's observation about hypertext, we argue, is also applicable to online communication:

> Meaning-making from the multiple linguistic, audio, and symbolic visual graphics of hypertext means that the cyberspace navigator must draw on a range of knowledge about traditional and newly blended genres or representational conventions, cultural and symbolic codes, as well as linguistically coded and software-driven meanings. (Luke, 2000: 73)

Two examples will suffice here. Users need to understand how to use the modes of representation they have available (e.g. clickable icons or emoticons) and how these can best be combined. They also have to be aware of the kind of structure that the electronic medium imposes on the conversation. Most text-chat tools, for example, create particular conversational patterns because messages are not visible to all until they have been sent. Intervening contributions displayed by the system can intersperse new topics before previous ones are concluded, obliging participants to adopt new norms of conversational behaviour (Rösler, 2004: 61).

CMCL presents the additional challenge that learners struggling with meaning-making via multiple modes in a new medium must also operate with a set of linguistic representational resources where they only have limited proficiency. This increased demand may be considered a key obstacle to acquisition, although the opposite sometimes obtains: less proficient learners who are comfortable with the technology may be valuable interlocutors for peers who are better linguists but worse technologists.

3.3.3 Multiliteracies: their implications for teachers and designers and for institutions

If teachers want to follow a sociocultural approach and believe in critical literacy, affective, social and critical skills become crucial. Multiliteracies go beyond dealing with the technical aspect of the electronic medium and include engaging with others through the new technologies and using these creatively as well as critically. For example, by using a reflective instrument such as a diary, instructors have been able to formulate a critique of the design of the environment (Vetter, 2003), or of their own management of online groups (Lewis, 2006).

Quote 3.4   Instructor's diary entry after a synchronous audiographic session

Monday, 8 September 2003
... talking about 'chat', L [my co-instructor] says it would be a good idea to be able to save the contents of the chat box, as this is where she jots down linguistic remarks and corrections.
Tuesday, 9 September 2003
Ah, very timely! One of L's students has just shown her how she can save the chat. Cool ...

(Vetter, 2003, personal communication)

Designers are interested in developing software that offers affordances potentially able to create social presence in a virtual, 'disembodied' environment where interaction might be limited to one or two modes. For example, BuddySpace (http://kmi.open.ac.uk/projects/buddyspace/) is a synchronous chat tool with a graphic display that shows users a map, with green dots in the locations where connected participants are, and red dots for those not currently online. It would be a missed opportunity to teach a BuddySpace session by concentrating on the text chat and ignoring the affordances of this 'social' map. However, if all students are connecting in the same town, it makes little sense to choose this software for the lesson. In this sense, teachers have a dual responsibility: to select the most appropriate tool for the job and to make the most creative use of the affordances of the tool that they have chosen.

Institutions may be more or less supportive of CMCL users (see Barr, 2004, for a comparative study of the support provided by three different institutions in the adoption of CMCL). Among the institutional factors liable to inhibit the creativity that we have said is needed from teachers are the managerial and the cultural.

First, decisions based on economics and security may determine that an institution will restrict online activity to one platform and will prohibit the use of some software. SKYPE™, for example, is prohibited throughout the French education system for security reasons.

Second, cultural factors play a role. Warschauer observes that in written CMCL, 'the decentered, multimedia character of new electronic media facilitates reading and writing processes that are more democratic, learner-centered, holistic, and natural' (1999a: 11). If true, this applies even better to tools developed since then, such as blogs or wikis. Yet many institutions still follow a teacher-led agenda and countless students are more familiar with hierarchical and instructivist learning contexts, as Chaptal (2003) and O'Dowd (forthcoming) have shown in their critique of secondary education in the US, French and Spanish systems. While this persists, there will be little opportunity for teachers to exercise the responsibilities that we have ascribed to them, or to find out whether learners benefit from such 'democratic' and 'learner-centred' features. Indeed, how to use the tools available to learners critically and creatively is a key issue for CMCL and for education more generally, and needs further in-depth research.

3.4 Summary

In this chapter we have introduced some of the major concepts in the field of CMCL. *Mediation*, a feature of all learning, needs to be foregrounded in any examination of the learning process where computers are involved. One way of understanding the specificity of computer mediation is by studying its *affordances*, and a useful way of operationalising the concept of affordance is to start by identifying the *modes* involved in making up a *multimodal* environment, then to consider the possibilities that they afford the learner, both as single and as combined modes. Implications for the definition of multiliteracies have also been suggested.

Further reading

From CMCL

Donato and McCormick (1994). Donato and McCormick discuss language learning strategies within a socioculturally-based pedagogy, through a case study. It provides an interesting opportunity to understand, through classroom evidence, how a theoretical framework such as socioculturalism can be implemented in teaching.

From other fields

Barbot and Lancien (2003). A collection of papers (in French) devoted to exploring the notion of mediation by humans (*médiation*), by various media (*médiatisation*) and the interrelationship between the two.

Conole and Dyke (2004a, 2004b). In the first article, Conole and Dyke give their understanding of the nature and role of affordance in computer-mediated learning, and in the second they defend this position against a critique. The dialogic presentation of the second article makes it particularly lively. Their primary focus is on affordance in design, but communicative and collaborative learning are also addressed.

Hutchby (2001). Although computer-based communication is a small part of the focus, Hutchby's book is particularly valuable in the contribution it makes to redefining the concept of affordance for communicative contexts.

Kramsch (2004). Using the notion of ecology, this edited book focuses on the interaction of language learners with their environment. It does so in four parts: interrogating models and metaphors for talking about language research, social practice, institutional history, and classroom practice.

Lankshear et al. (1997). For readers interested in literacy, this book explores how, in an increasingly complex world, literacy practices are changing. The final part of the book deals with literacy and new technologies.

4
Lines of Enquiry into CMCL

In section 1.6 we considered the quality of CMCL research, presented through a selection of meta-studies, which revealed a polarisation of views between those who hold that CMCL should be researched quantitatively and those who advocate a qualitative approach. Now we focus on two lines of enquiry that have particular relevance for CMCL. The first, comparative research, can be approached from a quantitative or a qualitative point of view, but in either case it raises questions of validity. We look at the work of one critic who has clarified the conditions under which comparative research may yield valid results. The second line of enquiry includes two related though distinct methodologies: discourse analysis and conversation analysis. They belong to a predominantly qualitative paradigm and as such the generalisability of their findings is often queried. Here we examine them in some detail and suggest how they might contribute to an increased understanding of the phenomena of mediation and the ecology of learning online. Finally, as part of the ecology of computer-mediated learning, we turn to the new field identified in section 1.5, intercultural learning online, and outline methodological directions likely to help research it.

4.1   Issues in comparative research

From the beginnings of CMCL those researching its effects on learners have structured their thoughts about what was going on in this medium by relating it to their previous, face-to-face experience. One reason for thinking 'comparatively' may simply be that there is an intuitive element to comparing the tools of today with those of yesterday: after all, our society is full of talk about the computer's capacity to help do things better and faster than before, whether it be producing novels, preparing accounts or mixing sound tracks. Second, designers of electronic environments rely on metaphoric comparisons with face-to-face experiences when they name their creations 'fora', 'rooms', 'whiteboards' or 'postings', which can be a reason why virtual environments tend to be perceived as if they were merely a variation on the face-to-face world. For example if, when logging on to a MOO (multiple object-oriented environment), one of the authors sees a screen indication reading 'MN enters' rather than 'MN connects', she is likely to perceive herself as physically entering a space (see also Jones, 2004: 25). Another source of legitimation for comparative thinking has been the distance teaching sector: institutions such as the UK Open University have over the decades documented the impact of their decisions to exploit technology in their teaching by comparing learner feedback elicited from earlier, less technologised situations with later ones (Kirkwood and Price, 2005). Yet comparative thinking undertaken for scientific rather than metaphorical or documentary purposes has to be carried out under particular conditions. The criticisms that comparative research has encountered are explicitly discussed in Allum (2002), who describes work supporting the need for comparative research. The clearest case against such research is found in Herring (2004). Her topic is CMC, but her observations can be applied equally to CMCL, and as such her work has inspired much of this chapter.

Quote 4.1   Herring's cautionary remarks about questions in comparative research

[T]he question should be *answerable from the data selected for analysis*. For example, if only computer-mediated data are to be examined, the question should not ask whether CMC is better or worse than face-to-face communication along some dimension of comparison, since the CMC data cannot tell us anything directly about face-to-face communication.

…

[T]he researcher is setting herself up for difficulty if she asks questions such as … 'Does membership in virtual communities satisfy needs previously satisfied only in face-to-face communities?' … [This] question involves a comparison; it can only be answered if empirical evidence (gathered by comparable means) is available from both 'virtual communities' (presupposed to exist) and face-to-face communities.

(Herring, 2004: 346, original emphasis, 347–8)

The consequences of the conditions described by Herring are that the same researcher would need to have carried out both parts of the study, or that the methods applied in the face-to-face study would need to have been applied also to the computer-mediated data. The chances of being able to carry out comparative research in naturalistic settings are slim, since cases in which the same learners are talking about the same topics, for the same reason, both face-to-face and through CMCL are rare indeed. This is why, for certain types of data that are unlikely to surface under experimental control, such as evidence of socio-affective bonding or intercultural awareness, empirical comparison of face-to-face and online community is not the best method and it may be more helpful to look for 'interpretive' answers rather than strictly empirical ones.

In the following sections we offer a detailed scrutiny of discourse analysis and conversation analysis as they relate to the evaluation of the nature and quality of CMCL interaction. We contrast the two methods, before illustrating the kind of 'interpretive' research that CMCL researchers can carry out.

4.2 Discourse and conversation analysis

One observation on which all agree is that CMCL produces large quantities of interactional texts, and that the computer-based nature of CMCL activities allows these data to be captured with ease through digital recordings of the visual, aural and written traces of human interactions. Therefore research methods oriented to the close scrutiny of such traces, such as discourse analysis (DA), and of social interaction, such as conversation analysis (CA), are frequently adopted. Indeed CA has been developed to deal exclusively with conversations, and DA's interest in text extends to conversations in so far as they are understood as particular forms of text. We look at how these two methods compare and how can they apply to CMCL exchanges.

4.2.1 Contrasting DA and CA

In a paper which we use as the basis for Concept 4.1, Santacroce (2004) provides a useful account of the commonalities and differences between DA and CA.

Concept 4.1   Santacroce's contrasting of discourse and conversation analysis

Commonalities

- Both are interested in naturally occurring conversations (exclusively so in the case of CA).
- Both rely on the sequential nature of conversations and aim to specify the norms that regulate conversations.
- Both regard the underlying logic of conversations as based on interlocutors' actions: for DA these are speech acts; for CA, they relate to the management of conversational turn-taking.

Differences

- Scientific affiliation: whereas DA has a strong relationship with linguistics, CA's roots are in the sociology of human interaction.
- Methodology: both DA and CA look for compliance with, or violation of, principles (the way analysts approach these principles is summarised in Concepts 4.2 and 4.3). For DA, though, the question is whether the discourse co-constructed by conversationalists is well formed in reference to principles of discourse coherence; while for CA the focus is on how the participants discharge their responsibility to the principles of sequentiality, for example in the relevance or timeliness of their inputs.
- Epistemology: DA is hypothetico-deductive. That is, its aim is to construct models accounting for the make-up of texts (in this case, conversations). CA, on the other hand, is inductive. That is, it aims to arrive at generalisations through description, in the greatest possible detail, of as much interactional and contextual data as possible. For CA, meaning-making is always a local matter.

4.2.2   Discourse analysis for CMCL

Discourse encompasses the relationship between humans and language, as humans use language both to reflect and to shape their social environment. Discourse can be realised through spoken, written or signed language, but it can also incorporate visual, musical and other forms of non-linguistic meaning-making. For example, while a recorded explanation of the location of the nearest exits is being played to aeroplane passengers, a cabin crew member may stand in the aisle and point to both her left and her right. In such a situation meaning is being made via two semiotic systems, one linguistic, the other gestural. In contrast, if I ask my fellow passenger where the nearest escape route is, he may respond by speaking and pointing, but is are unlikely to do so by getting up, standing in the aisle and pointing to his left or right. He is also unlikely to use the words 'location' or 'located'. In both situations the meaning is being made via language and gesture, but it is difficult, without further contextual and ethnographic information, to predict what kind of discourse the fellow passenger's communication will involve. When delivered by the cabin crew, on the other hand, the verbal and gestural phenomena are easily recognisable as the recurrent coded signifiers of a (commercial and institutional) discourse, expressed in more than one mode in this case.

Based on Herring (2004), we summarise the ways in which DA captures such recurrences.

Concept 4.2   Principles of discourse analysis

1. Discourse exhibits recurrent patterns, produced consciously or unconsciously.
2. Because of the unconscious nature of discourse, direct observation of empirically detectable phenomena is a priority rather than techniques of self-report through speakaloud protocols, interviews or surveys, which may be called upon as well but should be used with precaution (see also section 14.3).
3. Basic goal: to identify patterns that are demonstrably present, but not immediately obvious to the casual observer or to participants themselves.
4. Discourse involves speaker choices, not conditioned by purely linguistic considerations, but reflecting cognitive and social factors. Hence discourse analysis can provide insights into non-linguistic, as well as linguistic, phenomena.
5. In online settings: computer-mediated discourse and the uses that are made of the technological features of systems are in a relation of affordance, i.e. they may shape each other[3] (see also section 3.2).

When DA is applied to CMCL data, the unit of analysis may be structural (e.g. the structure of conversational threads) or content-based (cognitive, sociocultural or socio-affective). But in all cases the aim is to understand interactive behaviour through the meaning-making strategies that learners deploy, using such semiotic systems as are available to them in the environment under scrutiny.

4.2.3   Conversation analysis for CMCL

Many CA analysts refer to the object of their study as 'talk-in-interaction', because not all interactions studied in the field are 'conversations' in the

---

[3] As an example of this reflexive relationship, consider Lamy and Goodfellow (1999) or Stockwell (2003) regarding the negative impact of certain types of discourse on the sustainability of asynchronous threads.

everyday sense of the word. In this book, however, we retain the word 'conversations', which we find characterises learning exchanges in L2 and is simpler to use. Wagner rightly insists on the fact that CA has a strong 'ethnomethodological heritage – interested in describing how social order is produced in interaction' (1996: 232). A further feature of CA is that it describes 'the orderliness of social interaction as it is accomplished by methods and procedures that participants share' (Gardner and Wagner, 2004: 4). In our case this social activity is 'learning'. Although CA arose out of sociological work on everyday interactions pioneered by Goffman (1967), focusing on turn-taking in ordinary conversations with Sacks, Schegloff and Jefferson (1974), it soon recognised learning as a social interaction, one of the earliest writers to apply CA to classroom contexts being McHoul (1978).

Concept 4.3   Principles of conversation analysis

1. Turns of speech alternate and interlink. The basic principle is known as 'conditional relevance': e.g. given a question, expect an answer; given an apology, expect an acknowledgement; given a topic, expect that it will be pursued.
2. An ideal moment, or 'transition-relevant place', exists at which a new turn can be initiated without breach of politeness.
3. Face-saving is always important in a conversation, because conversants are in a relationship of perpetually converging and diverging interests with their fellow conversants. You may try to save your own face or to protect others'. It is also possible to threaten one's own or others' face for strategic reasons.
4. Any conversant is subjected to tensions between the need to talk in a way that is coherent with his or her world of reference and the need to maintain an acceptable relationship with the interlocutor. Preambles and precautions reveal the conversant's underlying beliefs, which s/he recognises need justifying when they are questioned.

These rules are all implicit and are just as revealing in the breach as in the observance.

As a method originally designed for dealing with everyday, face-to-face talk, can CA be harnessed in the study of talk that deviates from this starting point in two ways, that is, by being electronically mediated and by involving non-native speakers? To take the question of mediation first, CA's appropriateness for analysing online conversations is demonstrated in a paper on turn-taking in written chat in L1 (Garcia and Jacobs, 1999). In a departure from most of the contemporary literature, Garcia and Jacobs show that CA is applicable, providing it is customised specifically for this type of data, in this case through video-ing the screen of each learner. Regarding the question of L2 talk, although Wagner (1996) critiques the use of CA in the analysis of non-native speaker data, his arguments are convincingly countered by Seedhouse (1998, 2004). Seedhouse illustrates why he finds no reason not to apply CA to FLI (foreign language interaction) 'in the same way as to free conversation and to institutional talk' (1998, 101), acknowl-edging, however, that the non-native element of the talk makes the analysis 'even more difficult and time-consuming than usual'. This dif-ficulty may partly explain why fully developed CA descriptions of CMCL exchanges are still rare. Another part of the explanation may relate to the processes of CA research, and indeed of DA-based work too, which are complex due to the multimodal nature of the data (see chapter 16).

4.2.4   What can be researched through discourse/ conversation analysis in CMCL?

To see the kind of research questions that can be addressed through these methods, we reviewed studies that explicitly mentioned discourse or conversation analysis as part of their methodology. Unfortunately, CMCL researchers do not always clarify their relationship with these methods. Some claim to be using them when close interrogation of their procedures reveals no recognisable link with either. Others apply the methods yet fail to say so. We have selected studies with a declared dis-course or conversation analytical methodology, and we list them in Table 4.1 to show some of their applications to CMCL.

Clearly, a great diversity of topics can be addressed through DA- or CA-based methods. In section 4.2.5 we ask whether they are also able to accommodate a diversity of data types, such as are produced in the new multimodal environments.

4.2.5   Multimodal CMCL: beyond the limits of discourse and conversation analysis

To make meaning online, originators of messages may use a variety of semiotic systems. For example, they may leave a MOO by typing 'bye' (linguistic system), by clicking a button labelled 'Disconnect' (they are then using language, though not producing any themselves) or by clicking an icon (iconic system). From the point of view of the receivers, there is also a question of how the meaning (i.e. 'MN has dis-connected') is displayed: Is it received as a text-tag, or as an icon? The

Table 4.1   Applications of discourse and conversation analysis in CMCL research

Research areas	Studies	CA or DA
Resarching learner language		
Grammatical accuracy	Pellettieri (2000)	DA
Pragmatic functions and syntactic complexity	Sotillo (2000)	DA (some influence of CA)
Analysis of discourse (among chat participants)	Williams (2003)	DA
Individual differences in working memory and oral proficiency	Payne and Ross (2005)	DA (some influence of CA)
Researching communication		
Communication strategies to fill comprehension gaps	Lee (2002a)	DA
Patterns in native–non native chat	Negretti (1999)	CA
Learner engagement with native speakers	Schwienhorst (2004)	
Researching intercultural issues		
Intercultural competence	Belz (2003)	DA
Politeness and style shifting in different cultures	Davis and Thiede (2000)	DA
Researching affordances		
Techno-literacy, impact of the medium on ways that learners interact	Simpson (2005)	DA
Researching pedagogy		
Teacher training and teachers' cultures	Meskill et al. (2002)	DA
Critical pedagogy (power and equality in learning online)	Meskill (2005)	DA

former happens if their screen shows a message such as 'MN has left the room', and the latter if what they see is MN's name greyed out. In multimodal environments incorporating sound and the full possibilities of graphic user interface technology, the combinatory possibilities can be numerous and the semiotics of the communication can be complex.

Consequently, understanding the nature of conversational discourse in multimodal electronic settings is difficult without theoretical frameworks adapted to those environments. There is as yet so little published research on discourse in multimodal CMCL that we lack the basis for a debate about appropriate theoretical frameworks, but we can set out some conditions that would need to be met. For example, such a framework would need to:

- recognise that discourse effects are not only created through linguistic interaction but also depend on what Kress and van Leeuwen (2001) call the conditions of production, distribution and design of the 'artefact' (in our case the virtual environment);
- theorise the interplay between discourse features and the materiality of the virtual environment, possibly through a notion such as Hutchby's (2001) 'communicative affordance', which specifies the mutual impact of the content of communication and the technology used;
- theorise social actions in terms of the sensory space in which they take place and the location as well as spatial characteristics of any written language involved.

In relation to the last, of interest might be Scollon and Scollon's construct of 'geosemiotics', which they describe as 'the study of meaning systems by which language is located in the material world. This includes not just the location of words on the page you are reading now but also the location of the book in your hands and your location as you stand or sit reading this' (2003: x–xi). In the case of CMCL such a theory of sensory space would help account for the way the user's body engages with the computer, as each environment has its own interface structure, livery and graphic coherence. For example, environments may be based on different metaphors and use different colour codes and compositional schemes (the screen in some virtual worlds looks like an academic campus and in others more like a fantasy landscape). They may have different sounds, creating different connotations (for example SKYPE™, when connecting, sounds like a cartoon). Oral interventions by participants may be strictly regulated (only one person can speak at once) or freer (two or three voices can be heard at once), which has consequences for the pragmatics of the experience (depending on whether background laughter or sighing is or is not heard by interlocutors). In other words, we are calling for a theory of multimodal mediated discourse, extending to the specific needs of CMCL such concepts as have been proposed outside the field education.

4.3 The ecology of online learning, interculturalism and identity research

This section has a speculative flavour because of the paucity within CMCL of research specifically devoted to identity. However, identity is linked to interculturalism, which is increasingly debated within our field (sections 1.5 and 5.4). Identity is a notion currently under reconstruction by writers in face-to-face language learning theory (Norton, 2000) as well as CMC writers (Goodfellow, 2004; Goodfellow and Hewling, 2005). Their project has a critical dimension:

> [t]he notion of 'culture' as an essential attribute of individuals and groups, owed to national or ethnic background ... is unhelpful to the project of understanding how diverse participants in virtual learning environments (VLEs) individually and jointly construct a culture of interaction. An alternative conceptualisation of culture in VLEs is proposed, which views online discussion as just one of the sites in which the culture of a VLE is negotiated. (Goodfellow and Hewling, 2005: 355)

What might be the role of such a 'site' in networked language learning? According to Lemke, '[w]hat else is an *identity* but the performance, verbally and nonverbally, of a possible constellation of attitudes, beliefs and values that has a recognizable coherence by the criteria of some community' (2002: 72; original emphasis). In so far as the concept of community is important to online learning (see Concept 2.5), Lemke's insights about community boundaries are useful as pointers to where CMCL identity research could go. He makes two points. One relates to the site of cultural development that is the classroom, and hints at the architectural ecosystem within which the functions of traditional classrooms are played out: 'The classroom is no different from anywhere else in our world of social artifacts. Its developmental input is there not only on the walls but in the very fact that there are walls' (2002: 75). In the second point he is concerned with cultural boundaries: 'few communities today insulate their members effectively from the subversive texts and values of other communities. Barriers between cultures and languages are weaker today; our loyalties to them are moderated by our multiple lives and lifestyles' (2002: 75).

We suggest that Lemke's position has this to say to CMCL research: first, that in order to understand how virtual classrooms function as development sites for the identities of their residents, the very fact that its only 'walls' are lines on a screen and that they may be permeable (Jones, 2004) or not (Fanderclai, 1995) constitute a legitimate and important topic for investigation. This view supports the call for more attention to the materiality of the CMCL environment (see also section 2.3.1).

Second, Lemke's view of identity as 'performance', his opposition to the conventional model of language learners as monoglots seeking to become ideally 'fluent' in an idealised L2, and his insistence that success in language learning should be seen in the light of 'our multiple lives and lifestyles', suggest another line of enquiry in which CMCL research would extend its current domain of investigation. From seeing identity online solely as a locus of personal development, where desirable evolutions such as increasing tolerance of otherness should occur, it would go on to interrogate the wider social context of an online world with 'weaker' cultural barriers in operation. In this wider enquiry the theme of the colonising influence of an historically predominantly Anglo-American Internet on online education would expand and integrate critical concerns coming from intercultural activities in CMCL. (On identity and performance from the learner's viewpoint, see also section 6.4.)

4.4 Summary

In this chapter we have asked which research methodologies are suited to CMCL. Against a background of choice between quantitative and qualitative approaches, we explored the limitations of comparative research, followed by the role played by discourse as an object of research in CMCL. We offered a definition of discourse, explaining how it can be evidenced across different communication modes. We introduced a form of discourse of special relevance to online communication: conversational discourse. We contrasted the methods for analysing online exchanges (discourse and conversation analysis) and discussed theoretical challenges posed by researching conversations in multimodal virtual environments. Finally, we opened up the perspective of a new line of enquiry to address emergent notions linked to online identity expression and formation.

Further reading

From CMCL

Beauvois (1992). Interesting as one of the earliest examples of exchanges being conceptualised as 'conversations'. The 'slow motion' metaphor has endured.

Egbert and Petrie (2005). An important book providing a rich and clear picture of research issues in CALL, from both a theoretical and a practical point of view. However, as the book's topics traverse the field (e.g. 'Metaphors that shape and guide CALL research'; 'Considering culture in CALL research'), the specific coverage of CMCL is sometimes difficult to locate.

Debski (2003). In his analysis of 91 research articles published between 1980 and 2000, Debski seeks to establish the validity, conceptual quality and other characteristics of the research. The article is an excellent complement to Levy (2000).

From other fields

Cameron (2002). Cameron offers an introduction to spoken discourse, focusing on the practical as well as the theoretical. The book approaches discourse analysis as a subject matter as well as a method.

Dresner (2006). Dresner discusses conversational multi-tasking and offers a theoretical basis for assessing the multi-tasking potential of different communication media, including cognitive and cultural affordances as well as constraints of multi-tasking.

Mazur (2004). Although this work is geared to conversation analysis, it contains much good advice which could be of use to those approaching network data with other analytical methods. It also provides an interesting historical context for the field of analysis of online talk.

Sherry (2000). An accessible synthesis with findings that remain relevant in spite of the date of publication. Whilst this article does not use the acronym CMC, some of the author's comments about CALL are clearly aimed at issues of communication. The perspective here is interesting in that does not so much address the effectiveness of CALL research but rather how the research designs and instruments of the CALL field are structuring it as a discipline.

Wooffitt (2005). This book usefully explores the distinctive characteristics of CA and DA, and their interrelationships. Written with face-to-face and telephone conversations in mind, this reading can be complemented, from the point of view of CMC, with Dresner (2006).

5
Teaching Online

Teachers and tasks facilitate and mediate learning and constitute an important part of online learning (see Figure 3.1). A report on a Europe-wide survey on the impact of ICT in teaching and learning foreign languages, commissioned in 2002 by the European Community Directorate General of Education and Culture, claims that a

> shift of paradigm is necessary in teacher/learner roles. Co-operative, collaborative procedures are called for to harness the wide range of possibilities the new media offer. Teachers are called upon to abandon traditional roles and act more as guides and mentors, exploring the new media themselves as learners and thus acting as role models for their learners. (Fitzpatrick and Davies, 2003: 4)

So how does the online teacher's role differ from that of a teacher in a face-to-face classroom? What skills does an online tutor need? In what ways can it be said that task design has to adapt to the online setting? And what effect does collaborative learning have on students constructing their knowledge? These aspects complement the socio-affective factors relating to the student experience, which are discussed in chapter 6.

5.1 Teachers' roles and skills

5.1.1 The teacher as facilitator

In sociocultural theory, the tutor is no longer seen as an instructor and transmitter of knowledge. Instead s/he is a participant in the learning process, facilitates interaction among learners and guides them through their learning. Thus the teacher becomes a facilitator, a role that Richards and Rodgers explain as follows: 'In his or her role as facilitator, the teacher must move around the class helping students and groups as needs arise' (2001: 199) – by interacting, teaching, refocusing, questioning, clarifying, supporting, expanding, celebrating and empathising.

This principle has been particularly strong in CMCL; early scholars of online learning such as Mason and Kaye (1989: 27) point to the 'medium's' inherent support of a learner-centred environment', with tutors who are 'meant to be facilitators and resource people, available to be consulted when needed'. Debski (1997: 48) asks for realignments in the language teaching and learning process, and one of these changes includes the conception of the teacher as 'a facilitator, an inseminator of ideas, and a force maintaining the proper level of motivation of students' – mirroring the idea of language learners who 'become responsible, reflective and creative agents, taking over some responsibility for the outcome of the course'.

If we apply this notion of facilitating to online teaching and look at its different facets, it becomes clear that the facilitator does not have one role but many (Dias, 1998). According to Goodyear et al. (2001), these are: process facilitator, adviser-counsellor, assessor, researcher, content facilitator, technologist designer, manager-administrator. The point is emphasised by Shield, Hauck and Hewer (2001), who refer to the 'administrative tutor' or 'manager of learning events'.

5.1.2　Skills for online tutors

So far, most research about skills for online tutors has been carried out in non-language contexts. For a long time research focused largely on technical and software-specific skills – dealing with ICT problems and limitations, for example – but there is now growing recognition that technical expertise is not sufficient: 'To be an effective online tutor, it is clearly not enough to know which buttons to press in order to send an e-mail or which HTML coding is required to insert an image on a web page' (Bennett and Marsh, 2002: 14). Other necessary skills are to 'identify the significant differences and similarities between face-to-face and online learning and teaching contexts' and to 'identify strategies and techniques to facilitate online learning and help students exploit the advantages in relation to both independent and collaborative learning' (Bennett and Marsh, 2002: 16). Salmon's (2003) work on e-moderating additionally points to the need for tutors to go through a gradual build-up of competences as an online course progresses.

For tutoring languages online, Hampel and Stickler (2005) propose a 'pyramid model' with seven skills levels (see Figure 5.1) which include

Figure 5.1 Skills pyramid
(Hampel and Stickler, 2005: 317)

technical expertise, knowledge of the affordances, socio-affective skills and subject knowledge. Lastly, teachers should learn to teach creatively and develop a personal (and personable) teaching style in an online medium that has fewer/different modes of communication compared to the more familiar face-to-face setting. The authors also argue for the importance of training to enable teachers to become competent users of the functions of the technology, fully aware of its affordances.

A detailed analysis of the tasks of a teacher using a particular online tool was carried out by Vetter (2004). As an action researcher she describes her own experience with teaching via an audiographic conferencing system and lists the multiple tasks the teacher has to perform before and during an online session.

Additionally the tutor has to fulfil requirements which go across the categories identified in Figure 5.2 below, e.g. welcoming latecomers, allocating turn of speech sensitively, praising volunteer spokespersons, etc.

Research shows that an important part of the facilitation of learning is the skill to encourage the bonding of the online group in order 'to ensure that learning is meaningful, socially based and supportive of cognitive outcomes' (McLoughlin and Oliver, 1999: 40). This is particularly important for language learning, with its focus on communication. Yet the difficulty of combining social and cognitive outcomes is shown in the context of a study of asynchronous text conferencing, where Lamy and Goodfellow compared two tutors with different tutoring

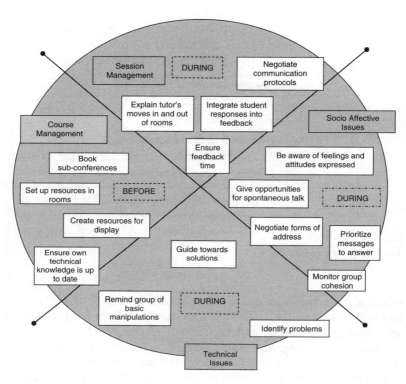

Figure 5.2 New tasks for the tutor
(Vetter, 2004: 123)

styles, 'one which places greater emphasis on the socio-affective needs of the students (social tutor) and the other giving a higher priority to students' reflection on syllabus content (cognitive tutor)' (1999: 475). While the social tutor's approach helped to foster learner–learner inter-action, and the cognitive tutor's style helped students raise their subject knowledge, neither of them managed to integrate both approaches.

5.2 Teaching online through collaboration, task-based and problem-based learning

5.2.1 Cooperation and collaboration

Cooperative or collaborative language learning is linked with the notion of the teacher as facilitator and the autonomy of the learner

(Macaro, 1997: 134; Richards and Rodgers, 2001: 192–201). This view of learning goes back to Vygotsky-inspired ideas of 'problem solving under adult guidance or in collaboration with more capable peers' (Vygotsky, 1978: 86). These ideas were taken up in Bayer's (1990) model of collaborative-apprenticeship learning, which, as Warschauer describes,

> emphasizes the use of expressive speech and writing, peer collabora-tion and meaningful problem-solving tasks. The teacher assists, not as a model but rather as a guide, while students collaborate to 'make connections between new ideas ... and prior knowledge', 'use lan-guage as a tool for learning', and develop 'language and thinking competencies' (7). (1997: 471)

We first explore what collaborative learning and cooperative learning mean. Roschelle and Teasley explain that there is 'a distinction between *collaborative* versus *cooperative* problem solving. Cooperative work is accomplished by the division of labor among participants, as an activity where each person is responsible for a portion of the problem solving', while collaboration may be seen 'as the mutual engagement of partici-pants in a coordinated effort to solve the problem together' (1995: 70). Oxford (1997) explores the implications of the differences, in a table that inspired Table 5.1.

According to Oxford's comparison, collaborative learning implies that students are in control of the learning to a greater degree than when they engage in cooperative learning. However, the concept of collaborative learning is also used more loosely:

> A definition of collaborative learning is when learners are encouraged to achieve common learning goals by working together rather than with the teacher and when they demonstrate that they value and respect each other's language input. Then, the teacher's role becomes one of facilitating these goals. (Macaro, 1997: 134)

Panitz (2001) lists the benefits of collaborative learning:

1 Academic benefits

- promoting critical thinking skills;
- involving students actively in the learning process;
- improved classroom results;
- modelling appropriate student problem-solving techniques;

- personalising large lectures;
- motivating students in specific curriculum.

2 Social benefits

- developing a social support system for students;
- building diversity understanding among students and staff;
- establishing a positive atmosphere for modelling and practising cooperation;
- developing learning communities.

3 Psychological benefits

- increasing students' self-esteem;
- reducing anxiety;
- developing positive attitudes towards teachers.

As early as 1990 Harasim linked collaborative learning to online learning. She believed that attributes of the online environment such as many-to-many communication or time independence would help 'to

Table 5.1 Conceptual comparisons among cooperative learning and collaborative learning

Aspects	Cooperative learning	Collaborative learning
Purpose	Enhances cognitive and social skills via a set of known techniques	Acculturates learners into knowledge communities
Degree of structure	High	Variable
Relationships	Individual is accountable group, and vice versa; teacher facilitates, but group is primary	Learner engages with 'more capable others' (teacher, advanced peers, etc.), who provide assistance and guidance
Prescriptiveness of activities	High	Low
Key terms	Positive interdependence, accountability, teamwork, roles, cooperative learning structures	Zone of proximal development, cognitive apprenticeships, acculturation, scaffolding, situated cognition, reflective inquiry, epistemology

(adapted from Oxford, 1997: 444)

explore the capabilities of online systems for facilitating educational collaboration and enhancing human thinking' (1990: 40). A whole field of practice and research has developed around the term computer-supported collaborative learning (CSCL), a field that has even generated its own academic journal (*International Journal of Computer-Supported Collaborative Learning*), although to date the language research community has not been well represented in it.

For collaborative learning to be successful, many critical factors relating to institutions as well as teachers and students have to be considered. Students, for example, need to possess or develop some degree of autonomy, and collaborative learning also requires group autonomy. This type of autonomy is 'the capacity of a group to manage itself on three levels: a socio-affective level (getting along with the others), a sociocognitive level (resolving problems together), and an organizational level (planning, monitoring, and evaluating work)' (Mangenot and Nissen, 2006: 604).

Collaborative settings or a collaborative course design do not guarantee collaboration, as Mangenot and Nissen (2006: 618–19) show in their investigation of an online language course. While the guidelines of the course 'insist[ed] on the necessity of interaction between students: discussing their interpretation of the documents, exchanging their ideas for the essay outline, and checking coherence between drafts ... there was little negotiation of meaning'. They blame this on the fact that the tutor was not allocated enough time to monitor the group and help the students develop new collaborative skills at a sociocognitive level. Institutional policies can also create obstacles – for example, the insistence on individual assessment rather than on work done collaboratively (see chapter 7).

Telecollaboration is one specific realisation of the idea of collaborative learning. It is usually arranged by linking students at different institutions and is now frequently used in online language learning as it allows one group of learners (who have language A as their L1 and B as their L2) to interact with another group (B as L1; A as L2). The aims are 'the development of foreign language (FL) linguistic competence and the facilitation of *intercultural* competence' (Belz, 2003: 68). Much recent research in this area has focused on the impact of social, institutional and cultural factors and on the examination of failed communication – see, for example, Thorne (2003) on the cultural embeddedness of tools, communicative genres and personal styles, or Belz's (2003) studies of what can prevent the development of intercultural competence in telecollaboration.

On the basis of a comprehensive review of the literature, O'Dowd and Ritter (2006: 629) propose an inventory of reasons for failed communication in telecollaborative projects, identifying issues at the socio-institutional, classroom and individual level. These are:

1 learner's current level of ICC (intercultural competence);
2 learner's motivation and expectations;
3 teacher–teacher relationship;
4 task design (thematic content, sequencing);
5 learning matching procedures;
6 local group dynamics;
7 pre-exchange briefing;
8 technology (tools, access);
9 general organisation of course of study;
10 prestige of language and culture.

This inventory is meant to help raise awareness among educators and encourage them to develop strategies to deal with such problems constructively. Rösler's (2004: 52–3) ten golden rules for the organisation of email projects with partner institutions in different countries (rules taken from the useful tips for teachers provided by Donath (2002) at http://www.englisch.schule.de/tipps_neu.htm#zehn), show the amount of organisational and facilitative involvement of the tutor.

So, as in other collaborative learning settings, tutor involvement and support in telecollaboration is crucial. Belz even goes as far as claiming that 'the importance (but not necessarily prominence) of the teacher and, ultimately, teacher education programs ... increases rather than diminishes in Internet-mediated intercultural foreign language education precisely because of the electronic nature of the discourse' (2003: 92).

5.2.2   Task-based learning

Task-based learning is a concept which has had a significant impact on language learning and teaching. While there are many different definitions of task (see Johnson, 2003), our focus is on communicative tasks, whose features were initially identified on the basis of the interaction hypothesis (Pica, Kanagy and Falodun, 1993) for face-to-face language learning. Ellis (2000: 200) outlines those features likely to have a positive effect on the quantity of meaning negotiation expected to take place.

Concept 5.1 Task features

- Information exchange required.
- Two-way information gap.
- Closed outcome.
- Non-familiar task.
- Human/ethical topic.
- Narrative discourse (vs. description/expository writing).
- Context-free, involving detailed information.

Most of these features are not media-specific – with the exception of the first two, where CMC has a particular contribution to make. Choosing from a variety of electronic tools, learners can exchange information via written and/or oral channels; and they can do so synchronously or asynchronously. This makes the interaction place- and time-independent. Exercises based on information exchanges and gaps are more authentic when students are not physically co-located.

The cognitive, psycholinguistic perspective has given rise to a definition of task that is widely advocated by researchers (e.g. Nunan, 1989: 10; Skehan, 1998a: 95, 1998b: 268; Yule, 1997) and which is summarised by Skehan.

Quote 5.1 Psycholinguistic definition of task

A task is an activity in which:

- meaning is primary;
- there is some communication problem to solve;
- there is some sort of relationship to comparable real-world activities;
- task completion has some priority;
- the assessment of the task is in terms of outcome.

(Skehan, 1998a: 95)

While works on tasks in language learning have long been concerned with psycholinguistic parameters (Candlin and Murphy, 1987; Bygate and Samuda, 2007), it has become apparent that the sociocultural dimension of learning cannot be ignored (see section 2.2). As Ellis (2003: 215) reminds us, the psycholinguistic and the sociocultural dimensions together 'mutually inform task-based instruction', the former being particularly useful for planning a task, and the latter important for 'improvisation', that is, the execution of the task in the classroom. So, when planning and implementing tasks, he suggests that in addition to the cognitive aspects the following sociocultural features need to be taken into account:

- verbal learning;
- private speech;
- zone of proximal development;
- scaffolding, collaborative dialogue and instructional conversations;
- motives, goals and operation of activities.

Meskill (1999) demonstrates how interpretation of task design can be usefully applied to our field, since CMCL offers settings within which learners can be presented with tasks comprising the necessary requirements (e.g. opportunities for exchanging information, discussing meaningful topics, resolving issues, working collaboratively to solve problems, constructing knowledge jointly, etc.). Meskill combines the two approaches to tasks informed by SLA and sociocultural theory into what she calls 'sociocollaborative learning tasks' and examines them in a CMCL setting. Apart from driving conceptual work, these tasks are active, participatory and meaning-centred, and value 'various perspectives – more than one way of seeing and solving a problem – and differing sorts of contributions on the part of learners [which] are particularly relevant for heterogeneous language classrooms representing a range of cultures and social educational strata' (Meskill, 1999: 145).

Quote 5.2   Sociocollaborative tasks in CMCL

These tasks:

- provide ample opportunities for differing perspectives and opinions, for controversy, disagreement, resolution, and consensus building;
- motivate active participation and interaction by having no one single answer or process to employ in accomplishing them;
- offer some form of problem-solving (something for which computers are particularly well suited);
- designate roles for individual learners and teams to take on as they engage in these processes, helping situate learners within a community of participants;
- and include a motivated awareness of the forms and functions of language used.

(Meskill, 1999: 146)

Meskill also recognises that 'the presence of the machine [i.e. the computer] inherently brings about major change in the structure and dynamics of discourse and activity' (1999: 154) and calls for careful task design. As Chapelle points out, we need to expand the scope of the basic approach to task theory – which has been developed directly from research on face-to-face tasks in the classroom – 'beyond the types of tasks that have been examined in the past to the types of CALL tasks of interest to teachers and learners today' (2003: 135).

A three-level process as a useful model for task development has been developed by Richards and Rodgers (2001: 20–31). It consists of *approach, design* and *procedure*, as follows:

- '*Approach* refers to theories about the nature of language and language learning' (2001: 20).
- '*Design* is the level of method analysis in which we consider (*a*) what the objectives of a method are; (*b*) ... the syllabus model the method incorporates; (*c*) the types of learning tasks and teaching activities the method advocates; (*d*) the roles of learners; (*e*) the roles of teachers; and (*f*) the role of instructional materials' (2001: 24).
- *Procedure* 'encompasses the actual moment-to-moment techniques, practices, and behaviours that operate in teaching a language according to a particular method. ... Procedure focuses on the way a method handles that presentation, practice, and feedback phases of teaching' (2001: 31).

Following Hampel (2006), we can apply the model to CMCL environments.

Concept 5.2   Model for task development for virtual classrooms

Approach

- Scrutinising theoretical frameworks and concepts for their ability to inform task design appropriately (e.g. ensuring that cognitive theories inform conversation-based tasks or that community building concepts inform simulation tasks).

Design

- Examining the triangular relationship between task type, tutor or student role and the affordances of the medium based on its materiality (see also Figure 3.1).

Concept 5.2   (Continued)

For example, how do these elements work together to determine the effectiveness of tasks conceived for use in threaded vs. unthreaded conferences? Or what can we say about the effectiveness of tasks designed for audiographic versus videoconferencing environments?

Procedure

- Thinking about how tasks can be orchestrated in the virtual classroom in order to foster interaction between learners and improve their communicative competence.
- Taking account of research to ensure more frequent participation, release more control to the students, enable collaborative work and a problem-solving approach, and negotiate certain pitfalls (e.g. issues of power online).

5.2.3   Problem-based learning

A further way of fostering collaboration and possibly empowering learners is through problem-based learning (PBL), which Oliver (2000) describes as characterised by a constructivist framework, one that 'encourages active construction of knowledge through personal inquiry, the use of problems to form disequilibrium and subsequent accommodating inquiry, as well as social negotiation and work with peers' (2000: 6). PBL is an instructional method that encourages students to work together and find solutions to real-world problems or scenarios. Giving students meaningful problem-solving tasks and allowing them to carry out an enquiry rather than formally instructing them makes learning an interactive, dynamic process (Rüschoff and Ritter, 2001). This involves reflecting on the problem, looking for appropriate resources and analysing them, and working out a solution, while the teacher plays a facilitating role.

Technology can provide a context for problem-based language learning (Tella, 1999: 114). Virtual worlds, for example, can offer rich, multimodal environments for engaging in more complex PBL tasks; written CMC affords a more reflective approach to discussions and can thus help students critically to analyse a set problem. In a special issue of *Distance Education* (2002) devoted to 'Studying Collaboration in Distributed PBL Environments', the usefulness of text-based CMC for PBL is discussed. In his comment on the collection of papers, Wertsch sums up the potential of PBL for a particular type of dialogue: meta-commenting in asynchronous written media.

Quote 5.3   Wertsch's view of problem-based learning

What computer-mediated PBL settings seem to provide ... are opportunities for this sort of extended, reflective commentary. These are opportunities that emerge due largely to the fact that the medium involved allows a single speaker/writer to hold the floor for as long as he/she likes and by the fact that one is not under the time constraints that characterize face-to-face verbal interaction. In other words, some crucial new properties of social communication – and presumably mental processes as well – have arisen with the use of this new form of mediation.

(Wertsch, 2002: 107)

While this may have potential for language learning, especially at higher levels, it is an area that requires more research, not just in the context of fora but also in connection with other tools (such as blogs) that are increasingly being used.

Problem-based learning is also often linked with the concept of open learning, which is addressed in section 6.3.

5.3   The teacher as reflective practitioner

Reflective practice brings together theoretical knowledge and experience (Schön, 1983). Two perspectives on the teacher as reflective practitioner are relevant in the context of computer-mediated language learning: reflecting on one's *teaching* via online tools; and reflecting on one's *online* teaching.

The sort of networking afforded by CMC can be used for reflecting on teaching as it facilitates teachers' exchanges with mentors or other teachers. Applying the concept of the reflective colleague, Russell and Cohen (1997) successfully used email to support one another with their teaching practice. Although the course that Cohen was teaching did not involve CMC, she and her co-author found that 'e-mail as a reflective dialogue has considerable potential for use in improving university instruction' (1997: 137). While email is fluid and spontaneous, it also gives an accurate and permanent record: 'E-mail was rapid, permitting responses within the same day or even a few hours. At the same time, it allowed time for thought and deep reflection, as we could mull over each other's words and ponder points or questions' (1997: 143).

Where the second dimension of reflectivity is concerned (reflecting on one's *online* teaching), several issues need to be considered. One is the initial training situation: trainee teachers need to experience the online classroom as their students will experience it, to reflect on it and also to observe more seasoned peers. The other is the continuing training situation, in which experienced online teachers can be seen as continual learners, trying to master fast-changing classroom tools and dealing with potentially similar anxieties about their use. Lewis (2006) provides an insightful account of his experience as a novice online tutor, setting out to teach in a multimodal online environment combining audio-graphic conferencing and *WebCT*™. He uses his capacity for self-directed professional development and employs three instruments – a teaching journal for reflection, a 'critical friend' for observation, a group of colleagues for discussions and feedback in an online forum – in order to develop his online language teaching skills systematically and 'inform pedagogic decisions in pursuit of a nondirective approach to teaching and learning' (2006: 581). Lewis's example shows that while some online tools such as email are being used more widely as teaching media and thus have become more 'interiorised', to use Ong's (1982) term, more recent tools – which at the time of writing included audiographic conferencing, podcasting and wikis – present new challenges. This can even result in reversed teacher–student roles (Lamy and Hassan, 2003: 51), with more technologically-able students helping other learners and even tutors to deal with the tool in question.

Research suggests that it is important for institutions to support and recognise reflective practice by, for example, providing tutors with mentors or reflective colleagues and giving them space where they can share their experience and find peer support. The importance of this for online tutors was emphasised by Hampel and Hauck (2004), who describe how training for a new CMCL tool was followed by the request for a tutor discussion area and how as a result a FirstClass™ conference was set up specifically for the 20 tutors teaching on this course. Hampel and Stickler (2005) found that tutors valued continued peer support or a mentoring system with more experienced colleagues helping novices, using, for example, an asynchronous written conference or dedicated webspace. However, as they point out, 'reflective practice also brings challenges as it requires a commitment to continuous self-development, needs time and training, and if it is done properly, can make practitioners question underlying fundamental values and assumptions' (2005: 323).

5.4   Summary

This chapter started by considering the roles that teachers play and the skills they need to develop in order to become competent users of the technology. Vetter's (2004) list of tasks that she performs before and during an online session illustrates these skills. The next section dealt with collaborative language learning and the fact that collaborative settings or a collaborative course design do not guarantee collaboration. As we have seen, tutor involvement and support remain crucial (e.g. in telecollaborative projects). Tasks were at the centre of the third section, where we showed the importance of taking account of the technological mediation when designing socio-collaborative tasks. A model for task development that consists of approach, design and procedure was therefore introduced. Finally, two perspectives on the teacher as reflective practitioner who reflects on his or her teaching via online tools, or who reflects on his or her online teaching, were examined.

Further reading

From CMCL

Dias (1998). Dias perceptively heralded the changes in and multiplicity of teacher (and learner) roles, that the literature of CMCL has pointed up in numerous studies since.

Furstenberg et al. (2001). The authors describe a project entitled *Cultura*, designed to develop the cross-cultural literacy of foreign language students. It uses the Web to access cultural knowledge and reveal hidden aspects of a foreign culture. One of the most influential projects in the field, *Cultura* helped to redefine the meaning of foreign language teaching in networked communication in the early 2000s.

Hauck and Stickler (2006). This special issue of *The Calico Journal*, entitled 'What Does it Take to Teach Online? Towards a Pedagogy for Online Language Teaching and Learning', combines the voices of practitioners and researchers and directs the research agenda towards further development of online pedagogy.

Opp-Beckman and Kieffer (2004). The authors develop a model for pairs of institutions to work collaboratively to deliver web-based online courses for the enhancement of language skills and cultural awareness. Issues addressed are: needs analysis, project planning, implementation and pitfalls.

From other fields

Ellis (2003). Despite its title, seemingly restricting its coverage to task-based learning and teaching, this book has a broad focus and gives an excellent and very detailed overview of general second language issues, including cognitive and sociocultural research approaches to SLA.

Shelley et al. (2006). This article examines the attributes and expertise required by tutors of languages in distance education, an area that – unlike face-to-face language teaching and the skills needed to deliver it – has been little researched.

6
Learner Experience

The focus in chapter 5 is teaching and the teacher, that is, tutor roles and skills, and teaching online through collaborative and task-based approaches. But what about the learners, whom Allwright and Hanks (2007) call 'key developing practitioners of learning'? What is their role in online communication and what is their experience? What are the socio-affective effects of technological mediation on the participants and on the way they interact with one another? The issues for discussion include student participation in online environments; motivation and anxiety; and the learners' sense of presence online and identity.

Most research has looked at written environments and has revealed both facilitative and inhibitory effects in terms of learner experience. Spears and Lea (1994) use the metaphors of panacea and panopticon (the latter a concept based on Jeremy Bentham's idea of a prison with a central tower from which all inmates can be watched at all times; see also Scollon and Scollon, 1995) to illustrate the contrasting views of CMC, thus juxtaposing the potential of CMC to empower and liberate the user with its risk of introducing surveillance and reinforcing unequal power relations between teachers and learners. Table 6.1 lists the potential benefits as well as problems for learners as identified in the literature and provides a useful structure for thinking about this contrast. It shows clearly that for many of the positive aspects of CMC there is a corresponding negative impact (see in particular the first six points in each list).

Table 6.1 shows that research findings about learner experience are ambiguous, often impressionistic and not necessarily transferable. The research effort to date has been unsystematic, except in the area of participation (see below). Research into socio-affective issues has only

Table 6.1 Positive and negative aspects of CMC for learners

Positive aspects

1 Equality of participation (written conferencing).
2 More turns (synchronous written environments vs. face-to-face classrooms).
3 Learner empowerment and autonomy; control of discourse by learners.
4 Time to reflect (asynchronous fora).
5 Less anxiety thanks to anonymity (written conferencing).
6 Greater opportunities for collaboration.
7 Authentic exchanges.
8 Creativity.

Negative aspects

1 Inequality of participation (written conferencing).
2 Lengthy monologues, flaming.
3 Limitation of learner empowerment and autonomy through greater control by tutor/institution.
4 Pressure to respond (e.g. prescribed number of contributions in asynchronous fora).
5 Increased performance anxiety (i.e. when speaking in synchronous audio environments).
6 Solitariness of collaborating at a distance.
7 Lack of paralinguistic cues and contextual deprivation can lead to misunderstandings, especially in written conferencing.
8 Information overload and techno-stress (multimodal conferencing).

recently started to emerge. In the rest of this chapter we explore these implications in more detail.

6.1 Learner participation

Early studies (e.g. Kern, 1995; Beauvois, 1998) showed that student participation in synchronous written conferencing was comparable to that in oral class discussion, resulting in more turns and more language produced (see chapter 4 for research on turns and section 1.4 for participation research). The fact that virtual environments allow learners to maintain much of their privacy and sometimes even grant them anonymity may also create greater equality for learners (e.g. Ortega, 1997; Kelm, 1998; Warschauer, 1998). Findings include the fact that in some cases shy learners are less deterred by appearance and social differences because these are not as prominent as in face-to-face settings (see e.g. Bump, 1990; Warschauer, Turbee and Roberts, 1996). Furthermore, CMCL has been welcomed because of its potential for allowing learners to take greater control of the discourse, as Chun points out.

Quote 6.1   Enhancing learner contributions

Computer-assisted class discussion (CACD) provides learners with the opportunity to generate and initiate different kinds of discourse, which in turn enhances their ability to express a greater variety of *functions* in different *contexts* as well as to play a greater role in managing the discourse, e.g. they feel freer to address questions to anyone or everyone in the class, to query the teacher form time to time, to suggest new topics or steer the discussion towards things they are interested in, to request more information or confirmation of something said by someone else, or to express thoughts or opinions that have not been explicitly solicited.

(Chun, 1994: 18)

Yet other studies show that computer mediation can amplify rather than hide personality differences and increase existing inequalities in participation, with dominating students even more likely to take over the interaction (e.g. Warschauer, 1997; Warschauer and Lepeintre, 1997; Lecourt, 1999). (For the impact of these issues on the assessment of participation, see section 7.1.3.)

These contrasting findings point to the fact that while CMCL can have a beneficial impact on learner experience, positive effects have not been sufficiently documented by research data. Realising the benefits of CMCL requires more than a virtual environment and includes other factors that need to be taken into account when collaborating in a CMC environment. Using a computer does not automatically increase or improve the quality of learner participation, as Hafner noted when discussing a 'commenting' function embedded in one of his projects: 'Students, when they commented, confined themselves to either thank you messages or expressing frustrations with technology/limitations of the materials' (2007, np). Issues such as anxiety, motivation and presence (which we explore below) also play a role, as well as adequate task design and the role of the tutor (see chapter 5).

6.2 Anxiety

We have found Csikszentmihalyi's concept of flow experience useful for understanding anxiety. A flow experience happens when participants are totally absorbed in an activity and forget everything around them. Csikszentmihalyi (1990) identifies a challenging activity that requires skills, clear goals and feedback, and a sense of control as preconditions that make such absorption possible.

Van Lier focuses on the first precondition and relates it to anxiety.

Quote 6.2   Conditions for flow experience

Preconditions for this state of flow are a perfect balance between available skills and challenges. Anxiety results from insufficient skills or insufficient challenges.

(van Lier, 1996: 106)

What causes anxiety in the context of language learning? One area for language learners is speaking (Hauck and Hurd, 2005). Gregersen and Horwitz call this 'communication apprehension', that is, 'an individual's discomfort in talking in front of others' (2002: 562). This is due to

> the mismatch between foreign language students' mature thoughts and their immature foreign or second language proficiency ... . The inability to express oneself fully or to understand what another person says can easily lead to frustration and apprehension given that the apprehensive communicator is aware that complete communication is not possible and may be troubled by this prospect. (Gregersen and Horwitz, 2000: 562)

In turn, anxiety can interfere with learning and performance (Horwitz, 2000: 256) by, for example, affecting listening comprehension and word production (Gardner, Tremblay and Masgoret, 1997). According to Dörnyei (2001b), one way of overcoming language anxiety is to develop learners' confidence.

Although research is currently critiquing the ideal of an all-encompassing native-speaker communicative competence (see Belz, 2002a; Kramsch, 2002; Davies 2003), language learners tend to have a perception that their skills are insufficient and that they have to make meaning with an incomplete set of resources. This may cause anxiety, affecting online learners and face-to-face learners alike. The question is what impact an online environment – where the computer adds a level of mediation compared to face-to-face interaction – has on anxiety. Learning a language in what is sometimes called a 'reduced-cue' environment means foregoing some of the modes of the face-to-face setting (e.g. vocal intonations and/or body language, depending on the setting; see also end of section 3.2.3 and Further Reading in Chapter 6). As well as reduction in cues, learners also have to deal with new functionalities and affordances that they may still be unfamiliar with, such as the threading structure in a discussion forum that allows for the organisation of messages, or avatars in virtual worlds that afford the users to take on different identities. This has cognitive as well as social and psychological consequences.

At a cognitive level, insufficient familiarity with the tools being used may cause techno-stress and cognitive overload, especially when technical challenges that overstretch learners are coupled with demanding linguistic work. It has been argued, for example, that in synchronous multimodal environments participants are under more pressure than in asynchronous environments which are dominated by one mode (Hampel et al., 2005). As learners' online multiliteracies skills continue to improve this may become less of a problem.

On the one hand, although asynchronous CMCL can cause stress because of the pressure to respond (e.g. in situations where participation is assessed), in terms of performance anxiety it affords learners the opportunity for reflection before responding, thereby reducing the pressure they may experience in spoken face-to-face communication. Written environments can also provide scaffolding by acting as fora for rehearsing oral language, thus potentially easing language anxiety (Payne and Whitney, 2002: 25; Roed, 2003: 170). The anonymity of a virtual environment can help overcome anxiety – learners may feel less inhibited when unable to see co-participants and their reactions (Benfield, 2000; Roed, 2003) and relationships may develop which are as deep as in face-to-face settings (Walther, 1996). Yet when it comes to the spoken mode, synchronous environments can be even more anxiety-inducing than face-to-face settings because of the lack of body language cues. There is some anecdotal evidence to support this claim. In Quote 6.3 one student's experience with interacting in the synchronous audiographic environment *Lyceum*, a network-based tool that allows for spoken and written communication as well as shared graphics, is described.

> ## Quote 6.3   Anxiety in audiographic conferencing
>
> *'[W]hen using Lyceum I would get more self-conscious about my language skills than when it was a face-to-face situation.'* He [the student] attributed this to the fact that *'my language skills were the only things being judged; I couldn't make myself sound better by smiling self-confidently and gesturing, and any pauses I made to think of vocabulary seemed incredibly long because there was just silence coming from every-one else; they could have been listening attentively, but I couldn't tell'.* However, he also added that this was only the case initially – *'it was soon apparent that everyone was accepting of whatever language levels their classmates may have been at.'*
>
> (Hampel et al., 2005: 16; emphases added)

Socially and psychologically, the geographical distance between participants has an impact. Some research has shown the greater 'alone-ness factor' (Shield, 2000), the 'contextual deprivation' and anonymity (see Peterson, 1997, for an overview of research) that can characterise online communication – at least in written CMC. A learner sitting alone in front of a screen rather than in the physical presence of others; all inter-action being mediated by a machine; and the fact that in CMC less can be assumed about other participants than in face-to-face contexts – these fac-tors can make it more difficult to develop and/or sustain motivation and lead to learner anxiety. This may be particularly difficult to pick up for tutors as well as for peers when there is no body language or any vocal presence. In extreme cases, these factors can lead some learners to pro-duce lengthy monologues or 'flaming', that is, students abusing one another (Beatty, 2003: 65). The likelihood of such harassment is lessened in teacher-moderated conferences, particularly synchronous ones.

The digital nature of CMCL means that all communication can poten-tially be traced and monitored. In the case of asynchronous written environments (fora or blogs), part of the pedagogical attraction is to be able to go back to previous postings and review them (see section 3.2.3). Yet the fact that, unlike in spoken face-to-face or telephone communi-cation, all communication is recorded and can be accessed by peers and teachers (possibly even a wider audience) can be inhibiting and anxiety-inducing. Getting something wrong in a spoken conversation is easily forgotten; whereas a mistake in a written conference is there for all to see. This sense of greater monitoring and control can increase anxiety and have detrimental effects on learning (see Lecourt, 1999, who explored this in a non-language context).

So how can anxiety be overcome and the sense of remoteness and anonymity that many users associate with virtual environments be minimised? How can learners be motivated to participate and communicate online, develop a sense of ownership of their learning and a sense of presence? And how can a feeling of flow develop, a feeling which two 'Traveler' users in Örnberg Berglund (2005: 9) describe as follows:

- 'It took me to another world and was a real adrenaline buzz. It was on my screen and I was conscious of it always, but I was definitely virtually gone from my usual habitat.'
- 'I'm always immersed. ... It doesn't matter that the environment is artificial. ... I think of the place as real.'

Motivation, control and autonomy are, we suggest, core contributors to such feelings of immersion and flow.

6.3   Motivation, learner control and autonomy

Motivation is the result of the interplay between intrinsic and extrinsic factors, between exploration and interest on the one hand, and external rewards on the other hand (Dörnyei, 1994, 2001a). Unless participation is required (e.g. by linking it to assessment), there are fewer extrinsic motivating factors in CMCL than in a face-to-face classroom. It is easier to remain in the background, especially in an asynchronous written environment where a teacher is not co-present to urge students to participate. Also, asynchronicity can result in participants not receiving immediate feedback from the teacher or their peers, and time-lags between postings can negatively affect learner motivation to communicate and interact (Benfield, 2000: np).

On the other hand, as Furstenberg (1997), Warschauer (1997), Tella (1999), Paramskis (1999) and O'Dowd (2006b) show, intrinsic motivation can be increased in CMCL by allowing learners to:

- write for a real audience (email exchanges or publishing work on the internet);
- develop useful technical skills;
- communicate with distant partners;
- work collaboratively;
- create projects that reflect their own interests;
- participate in authentic exchanges with peers and/or native speakers.

Debski concludes that 'a fuller integration of contemporary computer technology and foreign language education is most likely to take place in learning environments in which students can easily exercise their creativity, collaborate on projects of interest, engage in goal-driven activity, and combine learning a language with reflection about language learning strategies' (1997: 42). Learners can develop creativity by choosing the tools that suit them best (Mangenot and Nissen, 2006). Participating in meaningful activities such as developing academic research or writing skills, maintaining and promoting language and culture or producing an artefact such as a website (Warschauer, 2000) are other ways of increasing intrinsic learner motivation.

Intrinsic motivation is also linked to having control over the learning process, to autonomy and learner empowerment. In the field of offline language learning, Little (1991) has shown that autonomous learners tend to be those learners who are aware of the purpose of their learning programme, participate in the setting of learning goals, take initiatives in planning and executing learning activities, and regularly review their learning and evaluate its effectiveness (Dam, 1995). In the context of CMCL, Levy posits that 'the motivation derives from project goals and activities negotiated between students, or students and the teacher' (1998: 89).

Much has been hypothesised about the 'empowerment' of learners through the new media: 'it seems that modern information and communication technologies are not only tools but also create empowering learning environments that support constructivist concepts of learning' (Tella, 1999: 116; see also Engler, 2001). Especially in the area of written conferencing, students have been reported to be less passive and more autonomous – 'freed from the inhibitory effect of teacher presence' (Truscott and Morley, 2001: 22) – and to have more control over the context of language use (e.g. Warschauer, 1997; Kelm, 1998).

Before we examine this idea of empowering language students through online learning, let us briefly introduce a number of ways in which autonomy has been discussed in learning theories in general. At least three forms of learning are sometimes conflated with 'autonomous learning': open learning, distance learning and self-access learning. Of the three, open learning, which allows learners to set their own learning goals and assess their progress (White, 2003: 36), tends to be linked with autonomy. Distance learning (as opposed to open learning), as Candlin and Byrnes (1995) point out, can be associated with more conservative and instructivist pedagogies and they call for a move to greater learner-centredness. Self-access learning may simply consist in consumption of pre-prepared packages and will foster as much or as little autonomy as the instructional designers have enshrined in the materials. Caution also needs to be exercised when looking at the relationship of these three forms of learning to technology. None of them needs to involve electronic networking: all three were practised, albeit not with the current designations, long before the computer was invented. However, all three are commonly associated with electronic technologies, and one can see the potential of such technologies, for example, in distance learning. Whereas in the past there was little opportunity for collaborative learner-centred interaction (e.g. through problem-based learning; see section 5.2.3), the new media enable students to work together and co-construct knowledge even in a distance setting.

Yet we believe that there are two caveats to the proposition that there is an 'empowerment' of learners through CMCL. First, learner control is not intrinsically linked with the computer medium but can be achieved in other settings too. CMCL tasks presented in the research literature are often exploratory and constructivist, set up to avoid the instructivist IRF interaction pattern (i.e. consisting of initiation, response and feedback) that still characterises much face-to-face learning, thus artificially strengthening the apparent causal link between the medium and its pedagogical affordances. Second, there is often 'a discrepancy between learners' assumed autonomy and their actual skills' (Kötter, 2001: 332) and learners are not always given the opportunity to develop autonomy in the environment and context that they find themselves in. These restrictions may be due to the approach to learning that the tutor or the institution has adopted, or to the features of an online environment that gives the teacher more control than it does to the students (e.g. control over the turn-of-speech allocation functionalities of the system). Mason and Kaye have made the point eloquently in one of the earliest books on educational CMC.

Quote 6.4   Autonomy

The fact that this new paradigm [of online education] offers considerable choice and autonomy to the learner is irrelevant if the learner is not able to make informed choices about his/her learning requirements and to work independently of authority figures.

(Mason and Kaye, 1989: 25)

Having separated the notions of 'autonomy' and 'empowerment' from institutional forms of learning such as distance or self-access, as well as from forms of mediation such as electronic networking, how do we suggest that these two notions should be understood? Our answer is to situate them within the wider ambit of instructional design, specifically in reference to cooperative-collaborative learning, task-based learning and problem-based learning (see sections 5.2.1–5.2.3).

6.4   Presence and identity

We mentioned above that at one level anxiety can be linked to learners' sense of aloneness, contextual deprivation and anonymity in online environments. Allowing language learners to be situated in what Hutchby (2001: 1) calls an 'abstract form of co-presence' with others may thus not be sufficient to create an atmosphere conducive to interaction between participants (see also section 4.2.5 for a theory of sensory space online). So practitioners have tried to find ways to 're-create for online learners what is a crucial part of face-to-face communication: social presence (Rice, 1993; Gunawardena, 1995; Walther, 1996; Tu and McIsaac, 2002; Vogiazou et al., 2005)' (Heins, personal communication).

Concept 6.1   Social presence

Social presence clearly goes beyond mere knowledge and information exchange and has been defined as 'an awareness of a social situation in a group or community' (Prasolova-Forland and Divitni, 2003: 58), as 'a state of mind' that encompasses 'knowledge of others' existence, plans, motivations, intentions and attention' (Vogiazou et al., 2003: 2) or 'the measure of feeling of community'.

(Tu and McIsaac, 2002)

Some research has been carried out on how social presence is created in virtual environments (albeit little of it in the context of CMCL) but, once again, the work has come to contrasting conclusions. Thus some researchers argue that synchronous tools help to develop a sense of social presence and group cohesion more than asynchronous ones (Berge, 2000) or highlight the medium's potential to 'allow socialization and communication to take precedence over form' (Kelm, 1998: 19) and develop a shared social identity, that is, the feeling of belonging to a group. Yet other (non-CMCL) researchers argue that in terms of socialising, virtual environments require a different management of communication, as Mann explains.

Quote 6.5   Misunderstandings and conversational management in written conferences

[C]ontextualizing cues normally available in spoken discourse have been limited by the written discourse processes required. Furthermore, given the implicit nature of language ... the possibility for misunderstanding is greater and therefore the work required for 'conversational management' to mitigate this is even higher in this new environment. First meetings, early presentations of self, negotiations of learning community norms, and responses to contributors all have the potential for greater misunderstanding, all therefore become more significant and require greater effort to manage. ... A whole new communication process has to be learned. It is not simply a process of shifting from speaking and listening to reading and writing.

(Mann, 2004: 213)

Some studies have even suggested that online communication is so different from face-to-face interaction that students need to learn how to build social relationships and that socialisation in online learning is required, for example, through a face-to-face induction 'to provide enough social information with which to build a mental picture of other learners or tutors' (Nicol, Minty and Sinclair, 2002: 272).

CMC has implications not only for the social identity of a group but also for a learner's personal identity (which may revolve around a range of constituents such as gender, ethnicity or profession). The anonymity of many virtual environments and the functionalities of some applications allow learners to disguise their 20-year-old, white, female, middle-class self, for example, and take on a fictitious identity, or play with fictional, culturally shared identities such as 'Winnie the Pooh' (as we experienced with a British learner in one of our projects using instant messaging). Environments such as 'Traveler' or 'Active Worlds Educational Universe', where participants 'perform' as avatars, encourage such play; and home-pages and blogs promote the construction of a public persona. This idea of identity as 'performative' is relevant here: 'performative' describes something that is produced in social rituals (see Butler, 1990, on gender development) and can be played with. Performative identity is an area where some CMCL research is now available (Belz 2002a, forthcoming), but more is urgently needed.

It is equally important for CMCL to address this hitherto uncharted domain of 'identity as performance' by learners whose relationship to the English language is either problematic or nonexistent, or indeed harmonious but complicated by the fact that they may be part of an online group which is nationally heterogeneous, globally dispersed and collectively bi- or multilingual. (See also section 4.3 on identity and interculturalism.)

6.5 Summary

We started this chapter by looking at the beneficial as well as the inhibitory effects of CMCL on learner experience, focusing more closely on participation and anxiety as well as on motivation, learner control and autonomy. We found that CMCL's benefits in terms of increasing participation are largely potential – there also has to be a balance between learners' skills and the challenges they face, such as speaking anxiety, insufficient technological skills or cognitive overload. Learners also need to be motivated and have control over their learning. The concluding section dealt with issues that have surfaced more recently, issues such as social presence and identity.

Further reading

From CMCL

Coverdale-Jones (1998). A brief, early survey article in which Coverdale-Jones asks learners about the social aspects of CMCL. Their answers, at a time when the field was young, are still relevant. Interestingly, the reduced social content, so often a matter for concern in the literature, does not seem to be an issue for her respondents.

Peterson (1997). Although the overview this article provides is limited to written environments and deals with studies published in 1983–1986, it is a good starting point for considering the positive and negative aspects of learner experience of CMCL.

Sayers (1995). This very practical article advocates a mix of languages in the online classroom (if the pupils have them), using parents whenever possible to help enrich the polyglot quality of the atmosphere.

From other fields

Erlich, Erlich-Philip and Gal-Ezer (2005). Erlich et al. provide a good literature survey of interest to those wanting to build competencies for CMC participation.

Walther (1996). Walther discusses emotional issues of interpersonal use of CMC, with a particularly interesting focus on a phenomenon that has been described as the very opposite of the reduced-cue effect of CMC, that is, the hyperpersonal whereby affective contact may be more intense online than previously expected.

7
Assessment of CMCL

In maturing, CMCL has taken little account of assessment. Possibly, early research needed to attend to procedures, tasks, perceptions and a host of other aspects of online learning before it could speak reliably about the assessment of such a new form of learning, and the successive emergence of novel tools provided temptations for researchers to do more exploratory work rather than consolidate emergent knowledge.

Only a few CMCL authors have produced the kind of detailed account that has emanated from CMC, such as Hoffman's (1993) experience of using email as a feedback tool, or assessment of collaborative learning (Macdonald, 2003; Goodfellow and Hewling 2005). The language learning research community has instead tended to offer appraisals of computerised testing (Godwin-Jones 2001; Laurier 2003), studies of computer-based testing of specific skills (Goodfellow, Lamy and Jones, 2002; Coniam, 2006; Hafner, 2006) and studies of online feedback (Heift, 2001; Pujolà, 2001; Ros i Solé and Truman, 2005; Chiu and Savignon, 2006). A literature of assessment of online communication is yet to be developed. Fewer authors still have attempted state-of-the-art reports on assessment for CMCL: Chapelle and Douglas's (2006) synthesis of language assessment with technology devote their book to CALL rather than communicative skills testing. Yet the issue of assessing communication and collaboration online remains wide open, including questions such as: What is the role of summative assessment in networked learning? How should formative assessment be specified and carried out online? We begin by outlining what we mean by 'online assessment' before laying out the requirements of assessment schemes for CMCL.

7.1 Different understandings of 'online assessment'

When authors write about 'computer-aided assessment', they overwhelmingly mean tests that are computer-administered and/or computer-marked, such as multiple-choice questions (MCQs). However, this chapter assumes that efficient test-item delivery, automated mark-generation and computer-aided score management are not CMCL-specific and we concentrate instead on assessment of computer-mediated interactive language skills in its conceptual and experiential dimensions.

Assessing learning that has taken place *interactively* and *online* requires, we submit, an interactive online form of assessment delivery. Also, we believe that although the choice between assessment *of* learning and assessment *for* learning is in principle independent of the medium through which educational events are conducted, and although it is possible to use CMCL environments for any language pedagogy that one may deem appropriate, in practice there are affinities between online communicative learning and sociocultural theories (see section 2.2) and close links between CMCL and collaborative learning (see section 5.2). Teachers may indeed decide to assess the interactive quality of an individual's performance. Yet as the performance of a group is more than the sum of its individual members' performances, a strategy of individual assessment, if used alone, would remove the possibility of assessing the quality of an entire group's work. The focus on assessing the individual could thus be complemented by assessment of collaborative work, a form of evaluation that CMCL tools are particularly suited to enable, since they can support and in many cases easily track the actions of multiple participants.

What further issues arise from the human dimension of online assessment? To identify these we have found it helpful to set them against issues concerning automated assessment. These are brought together in Table 7.1.

Table 7.1 shows that human assessment online of discipline-based aspects (for language learning these include linguistic skills, the four skills of listening, speaking, reading and writing, as well as assessable cultural content) may have its problems. These include workload for assessors and assessment designers, the nature of criteria, constraints on test types and learner training. However, these problems are shared with other delivery modes and are not CMCL-specific. For this reason we have assumed that discipline-based considerations pervade the thinking of assessors at every stage of the construction of the assessment scheme, and in sections 7.1.2–7.1.4 we concentrate on the dimension that brings

Table 7.1 Issues with online assessment of interactive competence

Options when assessing online	Concerns with automated assessment online	Concerns with human assessment online
Formative vs summative		
Formative	Quality of automated feedback	Tutor-assessor training and workload
Summative	Fraud	
Assessing the product?		
Individual		
Collaborative		
Assessing the process?		
Individual	Limited to closed questions	Tutor-assessor training and workload
Collaborative		Assessment criteria
Discipline-based		
Quality of language and skills	Some language, skills and content can be tested via MCQs. Limited feedback possible.	Concerns are identical in offline and online assessment delivery modes
Quality of content (cultural and intercultural)		
Evaluator		
Self		Learner training
Peers	Learner training	Tutor-assessor training and workload
Tutor	n/a	

a specific challenge to CMCL practitioners: the assessment of collaborative learning.

Finally, given the emphasis on 'concerns', we should point out for balance that online assessment also has clear advantages over its offline variety, among which are:

- a good match between delivery modes (because if teaching is online, assessment should be online too, according to current consensus);
- easier reviewing and revision of test items owing to electronic storage and duplication facilities;

- easier re-usability of items, also owing to electronic facilities;
- administrative convenience;
- availability of permanent electronic traces of learner actions.

However, bearing in mind the CMCL community's reluctance to adopt online assessment, unresolved issues seem to predominate. We now give them a closer scrutiny.

7.1.1 Formative–summative and process–product assessment

A very clear reminder of the meaning of 'summative' and 'formative' assessment is provided by Hunt (2001).

Quote 7.1   Summative and formative assessment

Formative (continuous) assessment	An ongoing process of gathering information on the processes of learning, the extent of learning, and on strengths and weaknesses, which provides learners and tutors with information for future planning to meet an individual's needs; takes place during the course of teaching and *is essentially used to feed back into the teaching/learning process.*
Summative (terminal) assessment	Assessment which takes place at the end of a course of study or part of it. It measures learners' performance over that course or part of it; it provides information about how much learners have progressed and how effective a course has been.

(Hunt, 2001: 155; emphasis added)

The development of a culture of summative assessment online is inhibited by institutions' fears of fraud (Wolfe, 2001: 234). Cheating in interactive events may take the form of identity substitution (in synchronous chats and audio conferences) or of cut-and-paste plagiarism (in asynchronous forum debates where time permits the researching and unsourced importing of others' ideas). Atkinson reflects the technological determinism of many when he concludes that 'there are still a number of concerns about [the] reliability [of summative assessment online], but it is likely that ingenious solutions and new technologies will bring about a much greater degree of summative assessment than is currently possible' (2006: np). It has to be added that others (ourselves included, in our role as practitioners) are more interested in designing tasks that make fraud meaningless than in improving technology's fraud-prevention possibilities.

Currently, summative e-assessment is used as part of more comprehensive assessment schemes. With formative assessment, on the other hand, the issues relate to feedback, that is: the form that feedback takes (individual or collective); its timing (at which point in a synchronous event and how soon after an asynchronous posting); and its quality, particularly the subtlety of the descriptors used for discriminating levels of achievement.

Related to the distinction between formative and summative assessment is another difference, that between process and product-based learning. Observing how a learner performs a communicative language function such as telephoning for information, and assessing that performance against particular criteria specifying how that function should be carried out, is part of product-based evaluation. On the other hand, making an assessment based on observing the learner's linguistic and pragmatic behaviours whilst s/he is being taught to perform this function is part process-based evaluation. It is possible that an unsatisfactory performance in terms of product-based assessment may mask a great deal of progress in how the learner's behaviours have changed. This is why product and process-based assessment are complementary.

Summative assessment can be used to inform the institution or the learner about either product or process (although most of the time such information is about product). Formative assessment, on the other hand, is closely associated with process (see Table 7.1). What are the implications of online delivery for these assessment options?

7.1.2   Process and product of collaborative learning online

For process as for product assessment in CMC, the main issue is traceability. In computer-mediated settings it is easy to trace contributions and to identify their author(s). Here are two examples of the traceability of the process: with Learning Management Environments (e.g. WebCT™ or Blackboard™) the authorship of a collaborative product is traceable in the meta-data of individual components sent in by students; with some shared synchronous wordprocessors, student recasts appear in progress on screen, tagged with individual contributors' names.

Traceability has advantages for administrative purposes but also for pedagogical ones, as Macdonald (2003) explains in the context of written conferencing:

Quote 7.2   Process, product and collaboration online

[Online] assessment has a conspicuous advantage over the assessment of face to face collaboration, because the medium provides a written record of the interactions between students. ... This makes the process of collaboration more transparent, because a transcript of these conference messages can be used to judge both the group collaborative process, and the contribution of the individual to that process, thereby overcoming one of the traditional difficulties in implementing collaborative work fairly.

The other evidence that collaboration has taken place is the product, which may take the form of an essay or report, or perhaps a website. This can again reflect the individual contribution, for example each student may provide an individual critique of an online debate, perhaps assessed individually for the quality of reflection. Alternatively it can be a collaborative product, in which all students in the group are given the same mark, or it might have both individually and collaboratively assessed components.

(Macdonald, 2003: 378–9)

7.1.3   Process and the measurement of participation

Process-oriented assessment poses another problem: the assessment of participation, a three-faceted issue involving a quantitative and a qualitative dimension, and the phenomenon of 'read-only' participation, as we show next. Again, CMC rather than CMCL provides several examples for our discussion.

Assessment of student participation may be *quantitatively* measured against some requirement that learners should post a message a certain number of times per week (in asynchronous settings, blogs), meet particular wordcount standards (written, asynchronous or synchronous) or, less commonly, do some prescribed amount of speaking (synchronous audio). A problem with the quantitative approach to assessing interactive participation is that fewer postings and fewer words may in fact be associated with better interaction. Short, friendly comments, directly addressing individuals or groups, have been shown to trigger more replies than lengthy monologues. How, then, can discourse interactivity be appropriately measured? The question still has no answer, because although the interactivity of contributions can be captured – for example, by social networks analysts (Reffay and Chanier, 2003) – we have not located any examples of their being used as the basis for an evaluation of L2 interactive discourse competence.

Assessment of the *quality* of online interaction is slightly less difficult to find. However, there is little standardisation, leading to practitioners re-inventing criteria. For instance, in a scheme such as Anderson-Mejias', the value of content is judged according to whether messages demonstrate '(a) new idea, (b) restatement of previous idea with suggested change, (c) restatement of previous idea without change, (d) acceptance of idea(s) of others, (e) rejection of others' idea(s) with addition of a new idea, and (f) rejection of others' idea(s) without new idea' (2006: 26). Reviewing other qualitative assessment schemes, Goodfellow (2004) finds similar criteria, though some incorporate a higher degree of recognition of the dynamics and quality of the group's collaboration, for example, 'contribution to organisation of the group's activities', 'synthesis of new propositions', 'effective facilitation of the discussion', 'drawing on others' comments' (2004: 383).

To clarify our position regarding criteria, we believe criteria to be independent in principle of the delivery mode in which they operate. Instead, what differentiates an assessment scheme adapted to face-to-face delivery from one suited to CMC delivery is a twofold consideration: first, the relative weight accorded to criteria may vary (e.g. demonstration of success at deploying socio-affective skills needs to be given more weight when collaboration takes place at a distance); second, the assignment or task through which a candidate can demonstrate that s/he has met the prescribed learning outcome will vary, along with the delivery situation. An example of possible variation is that a criterion specifying 'contribution to task realisation' may be met in different ways: a pupil may satisfy the criterion in a classroom by assembling a three-dimensional physical object, or may satisfy it online by creating an electronic model of the real object.

However, as they stand, neither the criterion descriptors in Anderson-Meijas nor those itemised by Goodfellow meet our focus with sufficient precision, since they could apply equally to non-computer-mediated and computer-mediated communicative language assessment tasks. Perhaps Daradoumis, Xhafa and Pérez (2006), from a generic CMC perspective, offer a scheme that comes closer to reflecting the specificity of the online collaborative situation. The 'group functioning' and 'social support' criteria in particular are evidence that close attention has been paid to the social and affective dynamics of the group, an orientation associated – by consensus of many CMC studies – with the sustainability of online group activity. In Table 7.2 we reproduce this scheme in abridged form.

Table 7.2 Daradoumis et al.'s table of indicators of collaborative learning (abridged)

Collaborative indicators

Task performance

TP1	The students' individual and group problem-solving capabilities and learning outcomes ...
TP2	The students' contributing behaviour during task realisation (e.g. production function and use of active learning skills)
TP3	The students' individual and group ongoing (and final) performance in terms of self-evaluation

Group functioning

GF1	Active participation behaviour
GF2	Social grounding (e.g. well-balanced contributions and role-playing)
GF3	Active interaction or processing skills that monitor and facilitate the group's well-being function
GF4	Group processing (i.e. examine whether each member learnt how to interact and collaborate more effectively with his or her team-mates

Social support

SS1	Members' commitment to collaboration, joint learning and accomplishment of the common group goal
SS2	Level of peer involvement and its influential contribution to the involvement of the others
SS3	Members' contribution to the achievement of mutual trust
SS4	Members' motivational and emotional support to their peers
SS5	Participation and contribution to conflict resolution

We believe that these criterion descriptors are a step in the direction of a specification of online collaborative competences, yet they do not incorporate enough information about the evidence required if assessors are to consider that learners have met them, such as the discursive or linguistic form that text messages (e.g. in a written setting) or oral interventions (in a vocal one) might be expected to take.

This weakness leaves the criteria open to the criticism that they cannot be operationalised. For us, as for Daradoumis et al., the answer lies in adding statements that model student performance: as 'social grounding' skills (GF2 in the table) are associated with successful manipulation of objects in the online environment, learners might be shown a model of how to create a thread, how to alter the hierarchical position of a message in the thread or how to bring several messages together into a synthesis. We suggest that although this is a helpful approach, models with a finer grain may be needed, depending on local circumstances. The scheme in Table 7.2 exemplifies the considerable amount of work that remains to be done in qualitative assessment of online interaction, and points to the desirability of practitioner involvement in research (see Part III).

A final note about 'non-participation': in all the above contexts, little tolerance is shown of receptive-only participation. Yet silent reading or listening may be said to inform performance in acting as a form of scaffolding (Goodfellow, 2001). From the adoption of the pejorative term 'lurking' to refer to receptive behaviours, we conclude that assessment schemes overwhelmingly belong to a culture of measurement 'at the point of performance', rather than assessment of 'distance travelled'. To promote fairness in online assessment, further research on the role of apparent non-participation is needed.

7.1.4   Self-, peer and tutor-led assessment

Self-assessment is said to be one of the valuable affordances of networked delivery modes. Learners can very easily revisit their interactive performances. Yet although they may well do so informally, we have not discovered in CMCL documented examples of them doing so formally. It is also worth pointing out that Goodfellow and Hewling have identified a function of CMC-based self-assessment which is instigated to directly benefit institutions:

> Because of the practical difficulties of monitoring every student's contribution in an on-going discussion, the onus for demonstrating the value of their contributions may then put back onto the student, sometimes through requiring them to 'reflect' on [the] discussion at a later point, in order to provide evidence of their learning. (Goodfellow and Hewling, 2005: 359)

A better understanding of self- and peer assessment in networked learning is clearly needed, not only to develop appropriate assessment instruments, but also, if Goodfellow and Hewling are right, to decode the workings of institutional cultures.

Peer assessment is a desirable outcome of formal learning online in so far as it is in keeping with the form of teaching favoured on CMCL courses, as well as technically easy (ease of organisation of peer groups on 24/7 asynchronous and synchronous systems; ease of capture and circulation of completed tasks and discussion threads). Yet peer assessment faces two challenges. The first is common to online and offline settings and relates to the learners' inexperience in being assessed. At best this gap in learner skill means that time and resources need to be found for learner training. At worst, it means that learners may reject online assessment altogether, opting for more familiar forms of face-to-face testing. Macdonald (2004) illustrates the problem as well as a possible solution.

Quote 7.3   A view of peer assessment online in CMC

Peer review is a demanding task for undergraduates, because they need the confidence firstly to judge fellow students' work, and secondly to be able to give a critique without giving offence. However, with appropriate scaffolding to guide students, it is feasible, and has been demonstrated successfully with 200 students on a second level course [in Education] at the UK Open University.

(Macdonald, 2004: 224)

The second challenge is that the very format of online courses, often limited in duration to a term in an attempt to control the volume of work generated, militates against peer assessment. This is because, as Riel, Rhoads and Ellis point out in a study of peer review, 'there are reservations and issues involved in helping students develop the trust needed to work together effectively. ... To do this involves a process of reacculturation that is difficult to create in courses of limited duration' (2006: 143). The period deemed 'limited' by Riel et al. is 13 months – longer than many CMCL courses!

Tutor assessment of interactive tasks raises issues of e-literacy and workload for tutors, and of costs for their institution. For instance, in asynchronous, text-based work, the ability of tutors to exploit the technology in order to serve the needs of those being assessed is dependent on good e-literacy skills training. In synchronous settings where the teacher's time is limited, a key question is how to design and provide personal feedback without curtailing time for communicative outcomes. Also, enhanced support for learners before assignments as well as afterwards (feedback) is important, given that the remote setting carries a well-known drop-out risk. However, this has a high cost, as even the simplest test takes a considerable amount of time to develop. And providing high-quality feedback increases the workload considerably.

7.1.5   Mentoring and monitoring the e-assessors

Finally, the development of online assessing can be seen as a potential source of new thinking in tutor professional development. For example self- and peer assessment are also applicable to staff, particularly where institutional cultures support collaborative professional development. Whitelock (2006), one of the few writers on mentoring and monitoring of assessors in technology-mediated contexts, points out that in such research the long-term aim should be one that reconciled automation and personalisation of services to teachers, thus building

> a complete system that could be incorporated into a VLE [Virtual learning Environment] and would provide:
>
> - monitors with an environment that would enable them to focus on assignments that most need their attention
> - tutors [with] automated feedback on their marking, and contextualised staff development. (Whitelock, 2006: 275)

Realising this ambition is still remote, but research is ongoing in CMC if not yet in CMCL.

7.2   Designing assignments for CMCL

Given the options presented earlier (first column of Table 7.1), are some assignment design models particularly well suited to CMCL? Listing the possible electronic functions that can support CMC assessment schemes, Lam and McNaught warn that 'no one single e-learning design can employ all these possibilities' (2006: 214). They prefer to look at different combinations of design factors, a strategy we endorse as likely to be useful to assignment designers. Figure 7.1 marshalls the CMCL-specific assessment features of Table 7.1 into a representation of design choices which we also offer as an *aide-mémoire* for designers.

In Figure 7.1, the four main features of the assessment are show in capitals. Clockwise from top left, they are: formative vs. summative assessment; process-based vs. product-based assessment; self-, peer and tutor assessment; and individual and collaborative assessment. All test designers have to start with the question: Which one (two or three) of these features needs to inform my design? The position of these features on the outer circle of the figure symbolises the idea that as they work, designers should cyclically check that their design meets the requirements of, for example, formative peer individual assessment, or, to use another example, process-oriented collaborative tutor assessment. The second decision that designers have to make is: Which assessment output should my design produce? A list of example outputs appears at the top of the inner circle: so, for example, if the brief is to design an

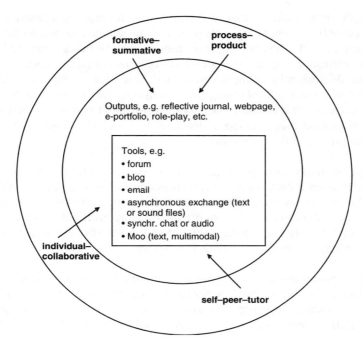

Figure 7.1 CMCL assignment design chart

individual assessment of process, an e-portfolio might be a suitable output. Last of the designer's decision should be the choice of tool (represented in a square in the middle of the figure), although sometimes there are institutional pressures on practitioners to prioritize this decision. Finally, some discipline-based issues pervade the entire area of assessment, such as how to assess intercultural learning.

7.3 The student's experience of CMC assessment

The student's experience may be approached from two perspectives: appropriacy of the experience in relation to the final result; and quality of the experience as a lived educational event. In the first category are included considerations of fairness, stress online, learner choice and appropriacy of the medium. In the second category come questions about collaboration inducing emotions and feelings of responsibility – sometimes even of guilt – in relation to others: e g. each learner feeling responsible for the success of the group. Many of these matters are generic, but some relate to specific tools. To illustrate fairness, students may be being 'protected' by their institutions against the potential unfairness of group marking of a forum discussion by being offered a mark based exclusively on an individual piece of work following that discussion. Yet having experienced this situation as a course tutor, Goodfellow (2001) points out that this results in their being denied feedback on how well they coped with the actual process of the discussion:

> giving marks for an essay (product) on the basis of how well the essay reflects on a discussion (process) is not the same as giving marks for the discussion itself. If the institutional policy on assessment rules out collaborative marking of discussion work, as it did in [this] case, then it is only the reflective essay that can actually be assessed, however important the discussion is considered to be. (2001: 79)

More positively, in self- or peer assessment schemes, the nurturing of assessing skills is a valuable personal development outcome as is reflection on others' learning. For example, easy electronic exchange of materials prompted one student, as reported by Macdonald (2004: 224), to say:

> One aspect of conferencing which I find useful ... is the exchange of actual pieces of work. With modern technology it is simple to attach a file of work to a conference message for people to look at and comment on ... It has proved very interesting and useful to be able to look at how other people have approached activities.

Finally, as students can benefit from looking at each other's work, so can tutors, see section 5.3.

7.4   Summary and future research needs

In this chapter we have discussed the need for clarity between generic assessment issues and those arising on the one hand from the mediation mode (online) and, on the other, from the conceptual model (interactive, collaborative). We have looked at online-specific problems, such as how to assess remote participation, discussed the design of interactive assignments for online delivery and identified some aspects of students' experience. This chapter has drawn many of its insights from CMC, an indication of how much work remains to be done to cater for the specific needs of language learners and their assessors. The gaps for CMCL researchers to pursue have been hinted at: feedback and error-correction policies; criterion specification; silent participation; self- and peer assessment; and e-literacy. Finally, although there is a growing interest in intercultural learning (see section 1.5), this is one area not featured at all in the chapter as – regrettably – no studies have yet documented its assessment.

Further reading

From CMCL

Chapelle and Douglas (2006). An excellent resource for those interested in a sound background to testing in CALL rather than CMCL, although the latter is touched on briefly, as are topics straddling both fields, such as issues of fairness to learners, and the testing of the oral language.

From other fields

Aljaafreh and Lantolf (1994). This article treats with great clarity the problem of making assessment into a learning experience.

Fulcher and Davidson (2007). Aiming to relate language testing practice, theory, ethics and philosophy, and designed with a variety of activities for the reader to try, Fulcher and Davidson provide a starting point for reflection whether the testing is off- or online.

McConnell (2006). Addressed to a continuing professional development (CPD) audience, this book focuses on the testing of collaborative learning online, including the principles underpinning such assessment, as well as design. One particular strength is its commitment to reflecting the 'view from the inside', i.e. how students experience collaborative assessment.

Shepard (2000). An article providing a broad historical view of the relationship between teaching, learning and assessment over the last two centuries, in offline education. The 'emergent paradigm' of socio-constructivist assessment is discussed in detail, and reflects the frameworks in chapter 2 of this book.

Weir (2005). Weir provides a framework for teachers and action researchers to construct and interpret language tests. It explains the nature of test validity and how to generate validity evidence. It covers all four skills, and includes a brief practical section on the testing of peer-to-peer interaction.

Part II
Research and Practice

Introduction

In Part II we ask what benefits have been derived from the application of research to practice, as well as whether and how empirical findings have fed back into research. The relationship between theory and practice is never one-way: instead it is reciprocal, and practical applications of theoretical concepts can lead to an interrogation and modification of the original theory. In some cases, it can also lead to a broadening of the theoretical base in which the research was originally grounded, as has happened, for example, with research on international telecollaborative projects, which has gone beyond the familiar issues of language learning and has opened up our field to wider psychosocial and semiotic considerations of identity and meaning-making online (sections 1.5, 4.3 and 6.4). The phenomenon which, following common usage, we call the feedback loop, is illuminated when such departures from the expected can be captured. In order to benefit from such insights, we have created five meta-studies, each based on two projects that have a tool (or in one case two comparable tools) in common. Thus chapters 8–12 are concerned with ten studies. In chapter 13, the final chapter of Part II, we explore technologies that may still be called 'emerging', in the sense that although they are in wide use in society at large and increasingly in educational contexts, little empirical research has been devoted to their use by the language teaching community. Having examined how the feedback loop functions (or in some cases fails to function), we make some suggestions about how research on emerging technologies could be approached.

We predicated the organisation of Part II on CMCL tools rather than other criteria, such as learning outcomes (e.g. linguistic, metalinguistic, intercultural), participant interaction (e.g. cooperative, collaborative, reflective) or task-types (e.g. problem-solving, project-based) or yet other factors (e.g. NS vs. NNS, tutor-led vs. peer-centred, etc.). Indeed our discussion of mediation (in chapter 3) and its representation in Figure 3.1 make it clear that we consider these criteria to be operationally interlinked. Yet in this chapter we chose technology as one of the many possible entries into the data. This choice does not mean that we endorse technological determinism – we do not – but technology belongs distinctively to CMCL and we use it in the following chapters as a device to assist us with contrasting different studies. As will be seen, much can be said about the functioning of the feedback loop in CMCL, and most of it is unrelated to technology as such.

8
Asynchronous Fora

8.1 Introduction

Asynchronous fora, whether linear or threaded, are the oldest tool in the panoply of CMCL. As they make low-tech demands on users, teachers and institutions were able to adopt them easily from an early stage in the development of the field. This long history has allowed a considerable body of research to grow up around asynchronous fora. Two groups of researchers, broadly aligned with the distinction we made in chapter 2 between cognitive and sociocultural pedagogies, have reported on them: those interested in the effectiveness of acquisition (attracted by the hypotheses that formal salience on screen and a leisurely response time might promote greater accuracy) and those more interested in the conditions for learning (attracted by the inherently social nature of networked learning). Given that fora have been in use for language learning for more than a decade, the two papers we have chosen to discuss below, dated 2002 and 2004 respectively, are newcomers. Although they use different populations (one involved pre-university classes while the other was carried out among graduate students), both aim to motivate their learners to debate by suggesting current socially relevant topics. Each study typifies a position in the distinct research directions signalled above. One of the articles is interested in investigating CMCL from the point of view of its effectiveness in promoting acquisition, and the other from the standpoint of CMCL's capacity to provide conditions for good learning. However, the outcome of a comparison between the two leads to new and very different questions.

8.2   Savignon and Roithmeier 2004

8.2.1   Which research frameworks informed the study?

The issues researched by Savignon and Roithmeier centre on three aspects of asynchronous forum-based language learning: text, collaboration and communication. The authors' view of text is inspired by Hallidayan text linguistics, specifically by its structuring principles of cohesion and coherence (Halliday and Hasan, 1976). Thus Savignon and Roithmeier are looking for evidence that 'the collected bulletin board postings on a single subject qualify as a cohesive, coherent text' (2004: 269). Their exploration of the second concept, collaboration, focuses on discourse as well: 'What discourse features can be identified that reflect participant engagement in terms of sustaining a collaborative dialogue?' (2004: 269). Third, the authors build on Savignon's (1983, 1997) theoretical work on classroom models of communicative competence to ask what potential benefit CMC brings to the development of communicative strategies by language learners. Although it is not explicitly developed in the article, the idea of linking a discourse-analytical question (on texts) to interaction (that is, to collaborating and communicating) seems to be that if a cohesive and coherent text can be evidenced across multiple authors' postings, then co-construction – through collaboration and communication – is considered proven. We return to this assumption in section 8.4. The starting point for Savignon and Roithmeier's reasoning is learner uptake of form, positioning the study within the more cognitive of the learning frameworks in chapter 2.

8.2.2   The setting

This project relates to a class of German students of English at a German secondary school (*Gymnasium*) who were eight years into their study of their L2. Working with them was a class of US students of German enrolled in their third year at a Midwestern high school. They discussed societal topics (the American Dream, the death penalty, drinking and driving, the Kosovo conflict) over a three-week period. They used English, a language which was of immediate interest to the German class, but was also an indirect motivator for the US students, who primarily wanted contact with German teenagers, and eventually aimed to have exchanges in German.

8.2.3 Insights from practical application

All three research questions received a positive answer. Uptake and peer-sharing of lexical items and of connotational knowledge were shown to have occurred 'implicitly or explicitly' (2004: 272). Thus the conclusion was that '[t]he cohesion and coherence of the postings for a single topic clearly qualify them as a text' (2004: 284). The issue of the sustaining of a collaborative dialogue received an equally positive answer, based on observations about both the use of strategies to mitigate potential conflict and the co-constructing of a coherent and cohesive text: '[t]hrough the incorporation of previously used lexical items, ideas, and even entire postings, participants show they were following the discussion' (2004: 284). Finally, Savignon and Roithmeier found numerous examples of communicative strategies. These consisted principally in 'resorting to the use of concessives or partial agreements followed by contrastive connectives' (2004: 274) to soften disagreements. Other communicative strategies were the restating – in this medium often via cutting and pasting – of others' ideas to introduce one's own differing interpretation, or involved the integration of peers' posts into one's own 'to express a personal view, while, at the same time, achieving global cohesion' (2004: 274).

8.2.4 Feedback loop to research and further practice

By using close Hallidayan text analysis of postings in order to establish a relationship between co-writing, sustained online participation and communicative strategies, Savignon and Roithmeier have contributed an analytical methodology that is novel in so far as its corpus of application is the entire exchange rather than individual postings. It would, they conclude, 'be presumptuous on the basis of a few weeks of CMC to make any claims concerning the acquisition of grammatical competence' (2004: 285). What they do strongly advocate is further exploration of discourse qualities (cohesion and coherence) in learner-posted CMC texts. They see this research as a constituent part of the theory of second language acquisition, due to the importance of the acquisition of discourse competence for learners. They are among the few CMCL researchers pursuing this line of enquiry. The merits of their approach have been debated in one of its aspects, that is, the use of strategies to mitigate potential conflict, such mitigation having been critiqued as an unhelpful pedagogical stance by Schneider and von der Emde (2006: 183). However, the more theoretical question of the well-foundedness of representing collaborative moves in terms of Hallidayan discourse analysis has not been debated. We suggest that such a debate would be useful.

8.3   Weasenforth, Biesenbach-Lucas and Meloni 2002

8.3.1   Which research frameworks informed the study?

The title of the paper, 'Realizing Constructivist Objectives through Collaborative Technologies: Threaded Discussions', clearly reflects the authors' aim, which is to determine under what conditions teachers may be '[r]ealizing constructivist objectives through collaborative technologies'. The authors use constructivist principles as a framework to evaluate the three-semester process of an implementation of threaded discussions to fulfil constructivist curricular goals. They use Bonk and Cunningham's (1998: 29) scheme, which is informed by the American Association of Psychology's taxonomy of principles expressing the nature of learner-centredness. The authors identify four overarching dimensions for researching learner-centred learning: cognitive-metacognitive, motivational-affective, developmental-social and individual learner differences. The study therefore fits well within the more sociocultural of the learning frameworks described in chapter 2.

8.3.2   The setting

A total of 52 advanced-level university ESL reading/writing students participated in the study over a three-semester period. All were international graduate students from various disciplines. Course requirements were explicitly oriented to participation patterns, for example:

> During semester 1, students were required to introduce a new thread each week and to participate in a total of 12 discussions about course content ... . In the first semester, students were required to post once per thread but were encouraged to post twice. (2002: 61)

The instructors' roles varied depending on the week and on learner proficiency levels, but all roles were predefined: observe, evaluate, provide ideas for discussion, join in discussion, model desired discourse. For example, one topic for discussion that instructors modelled for the students started:

> One argument often given for use of the death penalty is that it provides 'closure', a sense of relief, for the victims' family and friends.

How might the information in our first two readings be applied to this argument? (2002: 79)

8.3.3 Insights from practical application

In their discussion of findings, Weasenforth et al. concentrate on the role of the instructor and of the asynchronous medium before moving on to task design (with various assignments as task outcomes) and generalisability. The asynchronous nature of the threaded discussions, according to Weasenforth et al., 'makes this assignment particularly useful for the promotion of coherent discussion. The additional time available for reading and composing postings encourages reviewing and responding to classmates' arguments ...' (2002: 74). The asynchronous nature of the medium is also credited with enhancing the reflective learning style of quiet students, and with introducing flexibility through the possibility of extending assignment preparation time whenever necessary. One troublesome finding, according to the authors, is the low participation rate of some students. Overall the research question – can asynchronous fora help realise constructivist objectives? – can be answered in the affirmative, providing that forum activities are carefully integrated in the wider course and there is close monitoring from instructors.

8.3.4 Feedback loop to research and further practice

According to its authors, the contribution of this paper is that

> [w]hile previous studies have attested to the usefulness of asynchronous technologies in addressing constructivist principles ... these studies have focused largely on the social aspects of learning. The present study, however, provides a *broader* view of constructivist learning by examining not only social, but also cognitive, affective, and individual principles of learning. (2002: 59; emphasis added)

Their conclusions, highlighting as they do the need for integrated curriculum design and tutor support, bring a confirmation that principles of good constructivist practice expressed in CMC from the late 1990s onwards are indeed sound (for learning design, see Goodfellow et al.,1996; for tutor support, see Hara and Kling, 1999). That these principles should be reconfirmed for the field of language learning through a study that claims a wide scope is reassuring.

8.4   Conclusion

Returning to the perspective that has been opened up by presenting the two studies side by side, we make two points: on the one hand, we see asynchronous forum-based research as a field that has had the opportunity to develop sufficiently so that it has arrived at a consensus on acquisition; on the other, the field continues to be missing a theory of collaboration.

We make our first point based on the convergence of positions on acquisition from both articles. Both are careful not to say very much about acquisition. Weasenforth et al. have observed certain benefits for forum users in terms of the conditions in which the learning takes place, but claims of cognitive gain are not prominent in the study's findings. Savignon and Roithmeier, from their more cognitively-based starting point, make acquisition claims in a very guarded manner. It therefore appears that whatever one's position on the spectrum of learning frameworks, it is still necessary to remain extremely prudent about acquisition and effectiveness. Such caution, typical of many CMCL studies across the spectrum, could be interpreted as an acknowledgement that twelve years of CMCL research into acquisition have produced slim pickings. Leloup and Ponterio (2003) voice the concern of many practitioners when they conclude that there are problems with the CMCL research base itself:

> Researchers have yet to come to agreement on just what promotes and what hinders SLA. Much of the technology research base is centered on the investigation of computer use that facilitates or promotes those things that we believe aid language acquisition (e.g., interaction, target language input and output, acculturation, motivation) rather than on the measurement of outcomes. (Leloup and Ponterio, 2003: 1)

Yet we agree with Levy (2000), who takes pains to point out that the problem for CMCL (which he calls 'CMC-based CALL') is that:

> [f]or the CALL researcher, technology always makes a difference; the technology is never transparent or inconsequential. This is at the heart of what differentiates CALL from other cognate disciplines. CALL researchers are sensitive to the differential effects that specific technologies can exert and they endeavour to understand them through research. At the very least, in CALL research, the particular

technology in use is taken to affect the language produced, the learning and teaching strategies, the learner attitudes and the learning process. Beyond that, technology may affect the ultimate goals of learners, the nature of the learning environment, teacher education and what it means to be competent in a language. (Levy, 2000: 190)

Therefore, the fact that acquisition remains a largely unanswered question does not betray a shortcoming in CMCL research, but shows that the core of CMCL research concerns the mutual impact of tools and uses made of them by language learners, or, to use the phrase coined by Hutchby (2001; see section 3.2.1), the 'communicative affordances' of the technology for language learners. Whether questions of acquisition are secondary to this agenda and are likely to continue unanswered until the core knowledge is in place, or whether it is not in the nature of CMCL to answer them and whether they are in fact the proper object of applied linguistics, is a debate that it would be valuable to have.

The second point that we draw from the examination of the two articles on forum use relates to collaboration. Although both studies work within settings that have collaboration at their heart, neither problematises this construct. Weasenforth et al. examine CMCL data through a framework of learner-centred principles which, according to Bonk and Cunningham, 'are helpful but not enough' because they are 'far too broad and eclectic' (1998: 32). In particular, these authors envisage three views on collaborative technology: the learner-centred view is but one, to be complemented by a constructivist and a sociocultural view (1998: 44–5). Savignon and Roithmeier's treatment of collaboration is problematic in a different way, as they leave unexplained the relationship between text consistency and coherence, and collaboration. In fact, as we saw in section 5.2.1, the definition of collaboration has been the subject of much debate in offline language learning as well as in computer-supported collaborative learning (CSCL), which tends to study scientific learning, but not so far within CMCL itself, with the exceptions of Mangenot (2003) and Mangenot and Nissen (2006), who have strongly advocated closer links between the two domains (see also section 5.2.1).

In the present chapter, our comparison of two uses of a well-established CMCL tool has shown how the similarities in the settings (tool, tasks) influenced the research output negligibly, while the divergences in theoretical frameworks made a great impact on the findings. Each study has made a contribution to the understanding of our field. But equally, the process of comparing them brought into relief two of the challenges that persist in the field: the difficulties that CMCL researchers experience with the exploration of language acquisition; and CMCL's insufficiently developed links with neighbouring fields such as CSCL, which has, since the early 1990s, researched online group learning and has carried out much work, for instance in problematising collaboration, that could assist progress for the CMCL research community.

9
Synchronous Chat

9.1 Introduction

Chat programs were the first synchronous CMC tools made available for language learning and teaching. Outside of the educational context, chat is ubiquitously present in the learners' environment as instant messaging (IM), a tool that, in turn, is being used by some CMCL educators (Godwin-Jones, 2005). Because chat and IM can easily be logged, giving researchers instantaneous transcripts, they are convenient for examining written interaction. Although there are differences between them, for the purpose of this chapter we treat them as one.

Chat is written, but its synchronicity means that the language used is closer to oral discourse or, as Weininger and Shield put it, the language can be placed 'towards the "proximate" end of a continuum ranging from language of proximity to language of distance' (2003: 329). Therefore, it has been suggested that for contexts where the primary goal is acquisition, synchronous chat offers an ideal platform for rehearsing oral interaction. So it is not surprising that since the mid-1990s many studies on chatting have been carried out to test the effectiveness of synchronous written online communication for the negotiation of meaning, a concept hitherto used only in the context of oral interaction (see e.g. Chun, 1994; Ortega, 1997; Blake, 2000, see in this chapter; Pellettieri, 2000; Lee, 2002b; Payne and Whitney, 2002; Tudini, 2003).

Studies of co-construction of meaning among learners represent another main direction that research on synchronous chat has taken (Swaffar et al., 1998; Belz, 2001). More recently, researchers such as Thorne (2003; see in this chapter) have started to focus on the mediating role of synchronous communication tools and the effect this has on interaction between learners.

It therefore seemed interesting to us to look at the contribution of synchronous CMCL tools as exemplified by the earlier generation of researchers, using Blake (2000) as our starting point, and to contrast it with the role of these tools as seen by the interculturally-oriented researchers, exemplified by Thorne (2003).

9.2   Blake 2000

9.2.1   Which research frameworks informed the study?

Blake's study is based on interactionist SLA theory, which posits that 'the conditions for SLA are crucially enhanced by having L2 learners negotiate meaning with other speakers' (2000: 121). Researchers such as Gass (1997) and Long and Robinson (1998) believe that communication tasks can trigger this negotiation of meaning when linguistic problems arise and learners try to resolve these (see also section 2.1). Input incomprehensibility makes learners notice a gap, draws their attention to linguistic form and this is said to lead to modified input and increased comprehensibility through language modifications. Research in face-to-face settings has shown that particular tasks such as jigsaw, information-gap and decision-making tasks are particularly suitable as stimuli for the negotiation of meaning. The question for Blake is whether this also applies to an online environment. He formulates his goals as follows:

1 'to document that networked learner/learner discussions in Spanish would also produce language modifications such as those reported in the oral-based interactionist literature';
2 'to characterize linguistically those modifications'; and
3 'to test whether Pica, Kanagy and Falodun's (1993) predictions concerning the superiority of jigsaw and information-gap tasks also held for students involved in CMC' (2000: 122).

9.2.2   The setting

In his study, Blake set up networked learner–learner discussions via a synchronous chat program, 'Remote Technical Assistance', which offers users the following tools: a chat window (for one-to-one and group chat), a collaborative writing window, a shared whiteboard and a shared web browser. Fifty intermediate learners of Spanish took part in the study, working in dyads and carrying out a series of online task types: jigsaw, information-gap and decision-making. Blake also introduced one information-gap task which involved the dyads interacting with an unknown native speaker. Jigsaw tasks provide each partner with only half of the information needed to solve the communication task, and partners try to converge on a single outcome. Information-gap tasks assume that only one person holds the information, which the other partner must elicit. If this activity is repeated with reversed roles, it is called a two-way information-gap task.

9.2.3 Insights from practical application

Blake's main finding is that well-designed networked tasks 'promote learners to notice the gaps in their lexical interlanguage in a manner similar to what has been reported in the literature for oral learner/learner discussion' (2000: 132), with the advantage that CMCL allows the remote linking of learners. Regardless of the task type, 'the negotiations that arose in these networked exchanges tended to follow Varonis and Gass's (1985) typical schema' (2000: 125): trigger (the use by person A of a linguistic item unknown to person B), indicator (a signal by person B that there is a communication problem), response (or explanation by person A, trying to clarify) and reaction (acknowledging the help given).

Blake also found 'that jigsaw tasks appear to lead the way in promoting negotiations, as Pica, Kanagy and Falodun (1993) had previously predicted, but that information-gap tasks were not nearly as productive as a stimulus' (2000: 120). One jigsaw in particular created most opportunities for negotiation of meaning, for reasons of task design. Blake also asked what type of communication problems triggered most negotiation and found that lexical items did so, while grammatical items rarely generated negotiation of meaning.

9.2.4 Feedback loop to research and further practice

This study takes up Chapelle's (1997) call to use SLA theory as a basis for investigating CALL (see section 2.1). But it goes beyond Chapelle's triple agenda of (1) application of SLA theory to the design of CALL tasks, (2) execution of research that examines these, and (3) development of methodological research tools to allow more in-depth findings about their effectiveness. What Blake's study additionally does is demonstrate the limitations of this advice. As he acknowledges, while he has been able to show a 'positive impact of negotiations on vocabulary development' (2000: 133), chat negotiations have not allowed any conclusions to emerge concerning grammatical development. As Blake points out, it is difficult to prove 'that the negotiation of meaning also promotes the restructuring of the learner's linguistic system'.

Adding to this self-critique, we can also query Blake's methodology in its treatment of other aspects of language learning. This is an experimental study, conducted in a language laboratory, with students mostly sitting in one room while interacting with one another online. Would the outcomes have been different if the learners had not been co-located in a lab, but had participated at home from their own computers? Interestingly enough, the part of the study carried out with native speakers (who were unknown to the learners and were located in a different part of the campus) was also the one that produced least negotiation of meaning. Blake explains this by invoking the nature of the task as well as the inequality between native speakers and non-native speakers – the presence of the native speakers made it easier to keep the conversation going, but they were more in control of the conversation, possibly making learners embarrassed and nervous and less likely to acknowledge communication problems. Another reason may be the fact that participants did not know one another and had fewer shared classroom experiences. So this study is an example of an evaluation of the potential effectiveness of activity types rather than an evaluation of chat and its effectiveness. It shows that certain task types have the potential to create negotiation of meaning in face-to-face interaction as well as in chat settings, but that they do not automatically do so. Equally important to the assessment of effectiveness are questions such as who the participants are, where they are located, how they interact with one another and what social, cultural and institutional factors play a role in chat-based learning.

9.3   Thorne 2003

9.3.1   Which research frameworks informed the study?

As Thorne observes some ten years after the first studies on chat,

> there remains considerable debate, and some mystery, about the mediational affordances (e.g., the possibilities created by the relationships linking actor and object ...) of Internet communication tools and their correlation to linguistic and interpersonal dimensions of foreign language learning. (2003: 38)

Using data from three studies on telecollaborative exchanges, he set out to develop 'a conceptual framework for understanding how intercultural communication, mediated by cultural artefacts (i.e., Internet communication tools), creates compelling, problematic, and surprising conditions' (2003: 38) for the learner.

This conceptual framework is informed by a 'cultural-historical perspective of human communication and cognition', which emphasises 'the process whereby individuals modify, transform, and comprehend artefacts and environments' (2003: 39), including mediational artefacts such as chat tools. At the same time, this perspective helps to shed light on the connections between an individual's development and the surrounding social-material conditions. As Thorne explains, 'cultural-societal structures provide affordances and constraints that shape the development of specific forms of consciousness' (2003: 39). (For more on this mutual shaping through mediation, see section 3.1.)

9.3.2 The setting

Thorne bases his theoretical deliberations on three case studies of telecollaborative exchanges between US and French students. These exchanges were carried out via asynchronous and synchronous media and here we concentrate on the second and third case studies, where chat facilities were used. Both case studies involve students of French at Penn State University and engineering students at the Ecole Nationale Supérieure de Télécommunication de Bretagne.

In the second study, Thorne examines in detail the communication between a US and a French student, who started their dialogue with email but quickly moved to America Online Instant Messenger (IM). The data come from a post-semester interview with the US student reflecting on the exchange. The third case study focuses on the US students' perception of different tools, and in particular their preference for synchronous communication via IM over email. The data were collected through interviews as well as during one of three in-class synchronous chat sessions (using NetMeeting) between the key pals. For this, the US students were video-recorded in front of their computers, typing messages to their French counterparts and talking to one another and to the researcher who was co-present in the room.

9.3.3 Insights from practical application

Case study 2 counters the view of the psycholinguistic paradigm in SLA, which Thorne criticises for 'privileg[ing] the individual as an autonomous being, indirectly suggesting that real learning is *a-social* and unassisted' (2003: 51; original emphasis). In contrast, he shows that the learner at the centre of this study 'required the mediation of another person, specifically an age-peer who was willing to provide immediate and explicit linguistic feedback as part of a socially meaningful relationship' (2003: 51). He demonstrates that the student moved from object-regulation (through grammar texts) to other-regulation (through IM and email), allowing her to self-regulate her use of French. IM played an important role in this as it 'created the conditions for interpersonal communicative possibility that Csikszentmihalyi describes as "flow activity" (1990)' (2003: 53). (For more details on flow experience, see section 6.2.)

Case study 3 shows that although the exchange was supposed to facilitate intercultural communication between the students, the way it was set up did not take account of the 'cultures-of-use' of the communicative media. That is, it created a mismatch between how students use these tools in everyday life and how they were supposed to use them in the educational context. In order to talk to their peers outside class, the students tended to use IM rather than email. For them, email was a communication tool to be used with parents and teachers, that is, between power levels and generations, and therefore inappropriate for an exchange with peers in class and for building a relationship with them. One student hardly communicated at all by email with her key pal and only used the synchronous sessions to talk to him because she found the communication more dynamic and felt more comfortable when it was conducted in the latter medium.

The sociocultural perspective used by Thorne thus reveals 'the cultural embeddedness of Internet communication tools and the consequences of this embedding for communicative activity' (2003: 38). Tools such synchronous chat or IM (and even email) are closely tied in to the cultures-of-use, evolving from the way in which these tools mediate everyday communicative practice. Thorne shows how 'the mediational means available (e.g., IM versus e-mail) and their cultural-historical resonance for users, play a critical role in how and even if the communicative process and accompanying interpersonal relationships develop' (2003: 57). This is something that educators may want to take into consideration when integrating online tools into their courses.

9.3.4 Feedback loop to research and further practice

Thorne's is a theoretical as well as a practical paper. It draws on the findings from three case studies in order to support a conceptual framework of the use of CMCL in intercultural language learning, 'demonstrat[ing] that Internet communication tools are not neutral media. Rather, individual and collective experience is shown to influence the ways students engage in Internet-mediated communication with consequential outcomes for both the processes and products of language development' (2003: 1).

This is particularly relevant in CMCL, where there is often interplay in learning activities 'between students' non-academic identities and the discursively constructed institutional roles of the classroom' (2003: 4). If teachers want to use telecollaborative exchanges to build interpersonal relationships between learners, they need to ensure that the different cultures-of-use do not clash.

> For Internet-mediated interpersonal or hyper-personal relationships to develop, I suggest that certain minimum alignments of cultures-of-use are a necessary condition. In other words, the cultures-of-use of a communicative medium – its perceived existence and construction as a cultural tool – may differ interculturally just as communicative genres and personal style may differ interculturally ... Internet communication tools and their cultures-of-use, associated communicative genres, and for participant-actors, a shared orientation to activity, are necessary before substantive intercultural communication might develop. (2003: 24–5)

9.4 Conclusion

There is a sense in which our comparison of these two chat studies justifies the cautious statements we made about technological determinism in the Introduction to Part II. Indeed, the first conclusion that we reach is that Blake's and Thorne's distinct research orientations led them towards research outcomes formulated within the terms of the original respective frameworks, regardless of the fact that the tool under investigation happened to be synchronous chat. Such a conclusion confirms the final paragraph of section 8.4 where, from comparing two studies on asynchronous fora, we derived an understanding of the importance of frameworks rather than tools in determining outcomes.

However, to say that there is no role for tools to play would be reductionist. Thorne not only provides a corrective to potential reductionism, but takes the discussion one step further:

> Without endorsing technological determinism – the suggestion that technology determines human activity (an argument I counter in this article) – the structural properties of Internet communication tools have an effect on turn-taking and exchange structures ... . However, I wish to underscore and illustrate in the analysis to come that an artifact's materiality is conventional and takes its functional form from its histories of use in and across cultural practices. (2003: 40)

The turning point of Thorne's article, initially an exploration of how sociocultural conditions create 'varying qualities and quantities of participation in the intercultural partnerships' (2003: 38) of learners chatting across the Atlantic, is the active resistance to email in favour of IM by the US (and some of the French) participants. By re-orienting his research focus to the conceptualisation of cultures-of-use and their influence on learning, Thorne is able to return his reader to a level of reflection that goes beyond a limited interrogation of the affordances of chat or IM, towards a broad re-evaluation of educational philosophies, arguing that tools such as chat and IM present 'new challenges to the top-down organization of foreign language telecollaborative interaction (e.g., faculty-researchers making decisions about which CMC tools to use and for what communicative tasks)' (2003: 55).

In CMCL, for our knowledge to progress at all three levels – the individual's cognitive development, the group's sociocultural learning behaviours and the institution's offer to learners – we suggest that it is fruitful to make a variety of comparisons between technologies, between educational philosophies or between cultural groupings, and across these strands as well.

10
Multiple Object-oriented Environments

10.1 Introduction

MOOs (multiple object-oriented environments; see Quote 10.1 for examples of 'objects') have characteristics in common with synchronous text-based systems such as chat, which they combine with asynchronous written systems such as fora, but also have distinctive features, as Peterson (2004) explains.

Quote 10.1   What is a MOO?

[T]hese environments are designed around a hierarchy of user privileges that enables the creators of a MOO to structure the environment to meet the needs of a particular learner group. This feature enables users with appropriate privileges to utilize object-oriented programming (a unique element of MOOs) in order to create, manipulate and share multimedia objects and applications. ... A further novel aspect of MOOs is their spatial metaphor. In MOO learners can traverse a virtual space within a fully-featured virtual world that incorporates graphical maps and other navigation aids. ... MOOs adopt various learning metaphors such as for example, a virtual university. ... Typically MOOs contain numerous virtual rooms, linked together by entrances and exits. ... Some MOO environments contain numerous learning objects including virtual projectors, lecture spaces, notes, web pages and recording devices.

(Peterson, 2004: 40–4)

Therefore, when choosing two MOO-based projects, we were particularly interested in seeing how the language teaching community had exploited affordances specific to MOOs, in particular the system's unique customisability, allowing students to create objects that are personally meaningful and remain in existence even if no one is online. However the MOO literature yields little evidence to support the idea that there is detailed knowledge of MOO-in-use among the CMCL community even today. Shield, Weininger and Davies (1999), as well as Peterson (2001), have written detailed descriptions of the environment, and suggested task templates and elements of good practice, but were unable to avail themselves of sufficient empirical data to analyse the affordances of the medium in a systematic manner. In contrast, O'Rourke and Schwienhorst (2003), Schwienhorst (2004), O'Rourke (2005) and Warner (2004) mustered learner data, but kept the scope of their research constrained to the MOO feature that resembles synchronous text chat, so denying themselves the opportunity to study a MOO in its specific spatial object-oriented functionalities. One study which comes close to clarifying how the object-oriented nature of MOOs can help learning is the pioneering work of von der Emde, Schneider and Kötter: 'building rooms in the MOO is not just a pretend exercise, which students hand in and then forget. Instead, their rooms become part of the environment that the students construct and use for their language learning' (2001: 215). Unfortunately, empirical evidence for their claim is restricted to one room description composed by a student. No examples of object-oriented interaction are provided, and the bulk of the empirical material in the study is indistinguishable from data that might have been collected in chat projects.

Overall, therefore, the MOO literature divides into two categories: empirical studies with research foci on acquisition, which tend to concentrate on the 'chat' aspect of MOOs and particularly on tandem learning (see most of the authors listed above); and those that take the object-oriented nature of MOOs fully into account, more frequently providing anecdotal rather than systematically collected empirical linguistic data (e.g. Burdeau, 1997, for CMC; Shield, Weininger and Davies, 1999; Peterson, 2004, for CMCL, as well as the authors mentioned in the previous paragraph). The two studies that we chose are among the few recent papers that declare both a language-learning research focus and observation of learner response to the object-oriented, learner-extendable characteristics of the environment. The two have much in common, not least because they share some of the same researchers and institutions.

10.2 Kötter 2003

10.2.1 Which research frameworks informed the study?

Kötter's article explicitly calls on interactionist approaches to SLA, particularly the literature on negotiation of meaning, to help address the acquisition issues embodied in his research questions. But while much of his article and its findings are oriented to the interactionist framework, Kötter implicitly hints at Belz's (2002a) work on *language socialisation* when he speculates that:

> the game-like history of the MOO, its potential to provide a venue for role-playing activities and the relative anonymity that the users enjoy – depending, of course, on whether they are mere visitors or registered members of a language class – may not only prompt learners to experiment with unfamiliar structures, but that it may likewise stimulate them to explore (and exploit) the connotations of the language they are using and encountering in more depth than in a traditional classroom or a non-extendable chatroom. (2003: 150)

10.2.2 The setting

In autumn 1998, 14 German-speaking learners of English at the University of Münster and 15 US students of German at Vassar College met for twice-weekly interaction, for 75 minutes at a time, in a MOO where they collaborated in a total of eight teams of three or four students, to complete projects of their choice. For example, they created profiles about themselves and discussed cultural implications of this work with their partners. The German-speaking students were advanced learners of English, while the English speakers were intermediate-level learners of German. All teams were asked to present the outcomes of their work to the other groups during the final sessions of the exchange. Using log data as well as questionnaires sent to these cohorts, Kötter researched four questions:

1 How do students who meet in a MOO rather than in person deal with the apparent 'virtuality' of their encounters, that is, which (MOO-specific) tools and strategies do they employ to express themselves and exchange information?
2 How do the learners deal with utterances which they do not understand or situations in which they find it difficult to express themselves in their target language? Are the means they employ similar or different to those described in the literature on learner discourse between NSs and NNSs in settings other than the MOO?
3 How do the students exploit the fact that they meet as tandem learners, that is, how (often) do they request assistance, correct each other, help others through the provision of lexical assistance,

or scaffold their partners' tasks in other ways, including (deliberate) alternations between their native and target languages?

4 What evidence is there that the participants in this study improved their linguistic and metalinguistic competence and awareness as a result of their participation in the project? (2003: 146)

Of these questions the first directly addresses the specificity of the MOO, while the other three are more exclusively linguistic, have a bearing on negotiation of meaning and language awareness, and appear to be relevant to synchronous written environments rather than to MOOs as user-extendable worlds. How does Kötter deal with the first of his research issues?

10.2.3 Insights from practical application

Although the majority of his findings are acquisition-oriented and chat tool-based, the author offers us other insights which – had they received more elaboration – could have enriched our reflection on language socialisation through the medium of object-oriented, user-extendable MOOs. Many participants, he observes, had

> begun to make themselves at home in the MOO by creating their own rooms even before they met with their tandem partners for the first time. Some learners had fitted these rooms with objects ranging from a sofa or a carpet to a piano or a refrigerator, and several students had additionally composed elaborate descriptions of these purely text-based locales ... Equally important, many of the remarks that the students made to each other documented that they conceptualised the MOO as something with a spatial dimension. One learner commented upon arrival in her partner's room that it 'looks okay here,' while another stated that she preferred her peer's room to the MOO's entrance ... Moreover, many learners exploited the notion of space in the MOO by engaging with things they found in these rooms (e.g., Jack settles down in a comfortable chair) to create a pleasant atmosphere for their encounters even if they had no previous experience with MOOs. (2003: 152)

For Kötter, though, these socio-affective exchanges remain confined to a pre-sessional 'settling in' phase rather than becoming a part of the context for core learning opportunities available as part of the experience.

10.2.4 Feedback loop to research and further practice

Although Kötter provides much feedback for research on online negotiation of meaning, and on tandem learning, the data underpinning his findings again proceed from the text-chat-like facility of the MOO rather than from its learner-created dimension. The way that students 'accommodate' (to adapt Kötter's phrase (2003: 146)) to this dimension of the MOO environment is not interrogated further, nor is the relationship between their performance in the tandem exchanges or the success of their appropriation of the environment probed. In other words, the learners' use of the learner-extendable objects is treated as a preamble to the more concretely researchable instances of language acquisition. Kötter's first research question therefore remains insufficiently answered, at least for readers seeking an understanding of the affordances of MOOs.

10.3 Schneider and von der Emde 2005

10.3.1 Which research frameworks informed the study?

This work is underpinned by an approach to interculturalism that, following Bakhtin, the authors term the '*dialogic model*' (2005: 178) of learning, whereby 'any form of speech or writing – is not a self-unified system but the result and site of struggle, that is, conflict' (2005: 182). They are adamant in distancing themselves from communication as conventionally defined. Their strongly made point is that communicative pedagogues are wrong to insist on language students 'getting on' with others from different cultures by suppressing their desire to express opposition or disgust at foreign practises or habits that shock them. The European model of intercultural competence is praised as 'offering several important conceptual correctives' (2005: 180) to traditional communicative language teaching, notably because of its 'insistence on foregrounding and respecting cultural and individual differences among participants – in place of asking one group to dissimulate' (2005: 182), the latter remark being a reference to pedagogies that seek to deal with cultural conflict by avoiding contentious topics. In this project, gun-carrying stands as an emblematic topic for oppositional debate. Such opposition is not only 'tolerated' but managed as part of the ecology of the learner's developing argumentative and affective communication skills. Schneider and von der Emde provide no justification for their choice of a MOO (rather than another tool) as the vehicle for research on this pedagogy.

10.3.2 The setting

In autumn 2003, 11 German-speaking learners of English from the University of Münster and 14 US students of German from Vassar College met twice a week for 60 minutes at a time, for intercultural work in small groups in the same MOO as investigated five years earlier by Kötter. The central cultural content was articulated around a range of documents and activities, including (but not restricted to) synchronous discussions concerning two comparably shocking events: the shootings at a school in Columbine, Colorado, in 1999 and at a school in Erfurt, Germany in 2002.

10.3.3 Insights from practical application

Like other chat tools, MOOs have become *normalised* in their capacity as conduits for synchronous text exchanges, so that the part of Schneider and von der Emde's project called 'Exchanges' (i.e. synchronous chat discussions) is treated as a straightforward synchronous discussion, the dynamics of which are decoded via a form of content analysis. Schneider and von der Emde's 'Project Work', however, is a different matter. Based on student-created 'project rooms', the projects can claim to be specific to the MOO medium. How then do Schneider and von der Emde portray the activities carried out within these projects?

In their 'Project Work', students created spaces reflecting the discourse needs of their online community, that is, nearly all the groups built open-ended and interactive rooms in which to raise questions rather than give answers. These rooms 'depict all the conflicting perspectives on the various topics in order to encourage the visitors to the rooms to draw their own conclusions' (2005: 191). For instance, one group 'explicitly embedded the concept of conflict into the structure of their project by calling one of their rooms *Missverständnisse* "Misunderstandings"' (2005: 192) and posting the following explanation: 'This room has information about our discussions but it does not have any answers. Why? Because there aren't any concrete answers' (2005: 192). Schneider and von der Emde comment on such use of the environment, but their focus on cultural learning (rather than on educational technology) prevents them asking questions that might bring a better understanding of MOOs such as: How does MOO support students in matching the medium's functionality? By this we mean how does a functionality, such as enabling the creation of a room, support a discourse function such as inhibiting certain types of answers or facilitating reflection? (which happened when students opened a new room to discuss the conclusions reached in the room called *Missverständnisse*). How, in sum, would these affordances play out if the students did not have a MOO at their disposal but a chat, forum or blog?

In the second example we see one of the US groups working with the graphical affordance of the MOO. They created a room called *Media Effekt Raum* (Media Effect Space). In this room they placed an image of an enormous crowd in front of Erfurt Cathedral, actually taken during a German show of solidarity post-September 11, but erroneously interpreted by some 'visitors' to *Media Effekt Raum* as a memorial service for the Erfurt slaughter. From this pictorial stimulus a textual dialogue arose in which a rhetorical question with political-cultural implications was raised: Reversing the signification of the picture, how likely would it be for a US crowd to assemble in huge numbers to express solidarity with a non-US people's tragedy?

This is an instance of a group using MOO features to demonstrate critical engagement with the cultural agenda of the course. The Schneider study can be commended for touching on this, yet it falls short of problematising the relationship between discourse objectives (critical engagement) and MOO affordances (user-extendable spaces and graphics). What we claim needs to be asked is this: How does a complex architecture such as the MOO's, with its spatial metaphor, but also its graphical-cum-textual possibilities, function in support of the discursive realisations needed to enact situated learning about dialogical thinking, about tolerance of ambiguity or about critical engagement? And if there are task-design lessons to be learnt from the Vassar–Erfurt use of the MOO – and we think there may be – are they MOO-specific or are they applicable to other tools, and how are they formulated and disseminated?

10.3.4   Feedback loop to research and further practice

The working of Schneider and von Emde's dialogic model of intercultural learning is at the core of the study's findings. The feedback to practitioners includes warnings that 'a dialogic approach to online exchanges between language learners and NSs entails hefty risks' (2005: 198) and advice to expect that 'open-ended dialogues make it difficult to feel prepared ... because class sessions become driven by the kind of spontaneous interactions that take place between students (rather than objectives spelled out on a syllabus and revolving around a discrete and knowable text)' (2005: 199). Feedback is thus centred on the educational objectives of the research and includes no particular advice about the suitability of MOOs for achieving these aims, as distinct from other online environments.

10.4   Conclusion

In their conclusion Schneider and von der Emde admit that '[r]esearchers, including ourselves, are only now beginning to pursue the constraints and affordances for intercultural learning that inhere in particular forms of online media' (2005: 200). So after scrutinising the two studies on MOOs and noting Kötter's sole concentration on chat within MOO and Schneider and von der Emde's strong focus on pedagogy, rather than on the mediation of that pedagogy via a tool such as a MOO, we are led to the conclusion that insufficient notice has thus far been taken of the work of those who, as early as 2001, were trying to identify the specific affordances of the MOO medium. In the case of the Columbine–Erfurt study in particular, a radical rethink of pedagogy such as the authors advocate would be a particularly apposite object to hold up to the light of an innovative tool such as a MOO.

11
Audiographic Environments and Virtual Worlds

11.1 Introduction

Audiographic environments and virtual worlds are network-based tools for communication in real time and for the collaborative creation of text and graphics. The use of audiographic tools – integrating audio, text and graphics – in language learning started in the mid-1990s but has been slow to develop, possibly due to cost (unlike chat applications, most audiographic applications are not free). Virtual worlds are virtual reality programmes which range from immersive environments (with sound and touch sensors) to graphical spaces (with or without audio) and text-based environments (such as MOOs). Often they are open spaces, accessible by the general public as well as by those in a learning group. For the purposes of this chapter we concentrate on virtual worlds that offer more than the written mode.

We now explore two studies: an early piece by Erben (1999), who examined audiographics in an immersion setting for the learning of Japanese by trainee teachers in Australia; the other by Svensson (2003), one of the few practitioners advocating virtual worlds as suitable platforms for communicative and constructivist language learning. His article describes the use of one such virtual world with advanced students of English.

11.2 Erben 1999

11.2.1 Which research frameworks informed the study?

As Erben points out, his work takes place in a relatively unstructured theoretical environment:

> while audiographic technology used in immersion settings has the potential to enhance meaningful second language self-regulation as

well as to promote student-teachers' professional development, its creative applications have tended to be minimalised because it remains under-researched and under-theorised. (1999: 230)

He provides two frameworks for the study.

First, his pedagogical priorities are immersion and interaction, informed by a sociocultural framework which sees learning as 'inherent in activity that is culturally derived and mediated by various tools/texts'. Wanting to find out how students work together to become part of knowledge or speech communities, he asks the following research questions:

> How are linguistic and pedagogic processes (re)constructed in an immersion setting when instruction is networked through audio-graphics?' Specifically, (1) 'How is interaction mediated through audiographics and what implications does this have in terms of the professional development of student-teachers?' and (2) 'In what ways is student-teacher self-regulation facilitated or constrained in a context where instruction through language immersion in a teacher education setting takes place through networked audiographic technology?' (1999: 232)

Second, Erben's observations of learner behaviour and learner language in the audiographic environment are framed by his application of the SLA principles of input, interaction and output. He sees online interaction as a process carried out in such a way (i.e. remotely) that 'the inclusion of contextual information then becomes an issue for teachers and learners because there is no mutual sharing of the same spatial-temporal reference system' (1999: 237). This interpretation of online communication gives shape to Erben's findings, as we will see.

11.2.2 The setting

Erben's work is an experimental case study in an immersion context. It involves a BEd programme at Central Queensland University, in which up to 80 per cent of the curriculum is delivered through the medium of Japanese. In the first component of the study, student-teachers were observed and videotaped teaching face-to-face classes. They were then asked to reflect on their teaching in the light of the following question: 'In what ways is instruction in my immersion class constructed in terms of planning, organisation, communication, motivation and control?' (1999: 234). Their discussions were analysed and categorised.

The second part took place online through audiographics, with third-year student-teachers. Over the first three weeks they took part in twelve 30-minute sessions, working as a co-located group. For the next four weeks they were in different sites. The sessions were observed and videotaped, and the students and the teacher were interviewed.

11.2.3 Insights from practical application

Erben's commitment to sociocultural theory as well as principles of input, output and interaction produces an understanding of participants' behaviour couched in terms of 'reduction' and 'amplification' of the range of symbolic cues available to participants. Amplification and reduction refer to 'those classroom discursive practices which, because of the nature of the mediated interaction at a distance, participants need to modify ... in order to achieve the same effect as if the equivalent cue, sign or behaviour was produced in a face-to-face classroom' (1999: 237). For example, an amplification might be an instance of teachers having 'to increase question wait time due to the fact that delayed transmissions from site to site may occur' (1999: 238), while an example of reduction might be 'the loss of learning opportunities through such technical hiccups, where the connection between sites may freeze' (1999: 239). Initially, Erben found that in the online setting teacher-led activities were amplified; after a while, however, teacher control was reduced. Private classroom communication was generally amplified.

Another concept that Erben uses is that of 'reconstruction'. This refers to the development of modified discursive practices, as existing ones do not suffice to guarantee clarity of meaning and freedom from ambiguity in the new environment. An instance of a reconstructed sociocultural practice is that of 'bowing in the classroom', leading 'to the use of different verbal cues or picture icons' (1999: 240) in the online setting. Erben notes that 'as classroom participants adapted to the use of audiographics, instructional processes came to be increasingly reconstructed in ways which represented a substantive shift away from how these processes occurred in face-to-face immersion classrooms' (1999: 241).

From his mapping of reductions, amplifications and reconstructions to different components of the group experience in the virtual classroom – for example, moments of other-regulated activity and moments of self-regulated activity – Erben comes to the conclusion that an immersion education promotes

> the professional development of student teachers in the area of technoliteracy. Networked immersion education can be seen as a

mediated linguistic bath; one in which the student is far more active in regulating a range of pedagogical and linguistic processes compared with face-to-face immersion education contexts. (1999: 245)

11.2.4 Feedback loop to research and further practice

Erben's work was pioneering and almost ten years later remains one of the very few empirically-based studies of audiographic interaction. One of the present authors found his methodology initially helpful in her attempts to define the multimodal competences that audiographic environment users need to support their language learning (Lamy, 2004: 527). However, Erben's decision to peg online interactive behaviours to face-to-face ones, conceptualising them as greater ('amplified') or smaller ('reduced') than physical classroom practices, was found to obscure aspects of the meaning-making in the 2004 data. For example, in these data, unlike in Erben's, ambiguity did not trigger compensatory reductions or amplifications. Instead, the participants used a combination of linguistic and other modes to sustain a conversation that embraced ambiguity as a mechanism for creating humour, as was shown by focusing the study away from comparisons with face-to-face practices, and towards the interplay of interactions integral to the online situation (Lamy, 2004: 531). The 2004 research therefore benefited from the lead given by Erben's 1999 work not so much by applying his principles as by building on them.

11.3 Svensson 2003

11.3.1 Which research frameworks informed the study?

The starting point of Svensson's project was Bloom's taxonomy of cognitive objectives (knowledge, comprehension, application, analysis, synthesis, evaluation) as applied to the context of language learning. So the tasks ranged

from material-oriented tasks in the three sub-disciplines [linguistics, literature and culture] to more analytical tasks, and to a greater degree of collaboration, negotiation, and synthesis. Language skills come into all these areas: communication, negotiation, writing, virtual and real meeting, writing invitations tasks, explaining the world to visitors and so on. (2003: 133)

As the project developed, social, affective and psychomotor objectives, such as creativity or external participation, were included. The general framework of the project is thus constructivist, allowing students to build on their experience and to create knowledge actively. Learning is seen as a social activity and a holistic one, involving the body and the senses.

11.3.2 The setting

The project described was located at Sweden's Umeå University, at Humlab (http://blog.umlab.umu.se/), a state-of-the-art technology laboratory for humanities students and researchers. Entitled 'Cultural Simulation: Virtual Weddings and a Real Wedding of Linguistics, Literature and Cultural Studies', the project brought together different language subdisciplines. It involved advanced students of English whose coursework includes a major compulsory academic assignment, usually text-based and 20–25 pages long. The work within the project was required to be equivalent to an essay in terms of workload, but went far beyond the communication modes used in that genre, using linguistic modes (discursive and displayed text) as one part of the whole rather than as the principal mean of communication and combining them with images, movement and sound. Twenty-two students, the majority of whom did not have advanced computer experience, took part over three years.

The virtual space used – an environment called EVE (English Virtual Environment) – combines text and graphics and is run by the Active Worlds program. In order to participate, users choose to be represented by an avatar – for example, the body of a person or an animal, or even an object. Each person has to decide whether they are able to see this avatar or whether they take on a 'first-person perspective', looking out of the avatar's eyes. For the purposes of this project, students worked together in small teams to build a graphical world around a theme ('weddings' in year 1; 'the city' in year 2; 'monstrosity' in year 3), to link objects in their 'world' to webpages they created (containing, for example, essays) and to represent, often in non-linguistic form, the concepts relevant to their theme. In the end, the teams presented the 'world' to their peers and their teachers. So, while students still did some of the more traditional academic work (e.g. essay writing), this was integrated into a larger project which required collaboration, negotiation of decisions and allowed them to use their imagination in a multimedia environment.

11.3.3 Insights from practical application

In his article, Svensson presents some of the students' work (e.g. a project presentation and images from the 'worlds') and discusses it within a constructivist framework. The project is deliberately process- rather than product-oriented, and he shows how students collaborate and negotiate in building their 'world' and representing the central concepts, using images, concepts and sound (e.g. music). He stresses the importance of creativity and motivation, claiming that while text can be creative and dynamic, 'visual and auditory means of expression in distributed spaces have a very strong motivating and creative effect' (2003: 139). According to Svensson, such project work is particularly useful for language learning, which 'is about language, immersion in other cultures, communication, media, intercultural meetings and role-play, and virtual areas supply us with a place where all these can come together' (2003: 140).

11.3.4   Feedback loop to research and further practice

Svensson's study indicates how new virtual spaces can be used creatively for language learning purposes, thus challenging a text-based paradigm in university education. His research also shows how important it is to be aware of the affordances of the medium: 'We need to work with the medium, and think about how it can be used most suitably for our needs in the new context' (2003: 129). Rather than applying principles from face-to-face contexts or from written chat environments, Svensson believes that employing a particular medium for teaching and learning purposes implies taking advantage of the modes available within it: for example, if students are to look at a cultural concept such as 'the city', their work may be more appropriately expressed through the use of their own spatial designing of a city, which they can co-construct, than by traditional linguistic means.

While such projects open up exciting possibilities, they are still exceptional in educational contexts and almost invisible in the litera- ture. Svensson himself has published few theoretically-based accounts of the groundbreaking work described above, preferring to use Humlab to showcase such student productions. His work may be exceptional as his project was located in a state-of-the-art computer lab, funded by external resources, supported by educational specialists and computer support staff, and involved only a small number of students. How such well-resourced virtual environments can be usefully employed in more mainstream language learning contexts remains to be seen.

11.4   Conclusion

Erben faced the uncharted territory of audiographic learning in 1999, and in 2003 Svensson studied the then unexplored domain of virtual world-based learning. The two studies are both pioneering, but are in great contrast to each other. Erben has helped define unfamiliar phenomena in the online world by mapping them to the known reality of the physical classroom, thus creating a model that could be interpreted by later researchers as limiting, or as a default position from which alternative models could be constructed. Svensson, on the other hand, appears to have succeeded in emancipating his learners from the constraints of former ways of learning. Their creativity has produced collaborative virtual world objects that are inspirational examples of what can be achieved in virtual worlds. However, to date the learning achievements underlying these technological achievements have not been made available to the research community as theoretically-based accounts – with the exception that we have just explored.

More research is needed in both areas, research that takes up the different methodological challenges thrown down by Erben and Svensson, and investigates the interplay between different modes of communication (e.g. text and images, text and sound), as well as issues relating to virtual presence and to identity, regarding the shared social identity of a group as well as individual questions of identity, for example, in connection with the use of avatars in virtual worlds.

12
Videoconferencing

12.1 Introduction

We have chosen two videoconferencing studies which were published a decade apart, a period during which the technology that they reported changed radically. One involved a technician-controlled camera (Goodfellow et al., 1996) and the other learner-controlled webcams (O'Dowd, 2006a, 2006b). In the decade that separates them, some degree of what Bax (2003: 23) calls normalisation (the stage when the technology becomes invisible, embedded in everyday practice and hence 'normalised') has occurred and the videoconferencing tools have become more flexible.

O'Dowd examined the literature on CMCL videoconferencing published in the years between Goodfellow et al.'s and his own projects. He presents a mixed picture of the impact of seeing faces on screen. On the one hand, he found Buckett, Stringer and Datta (1999) reporting, against Goodfellow et al., that the medium functioned well as it provided a way of decoding reactions, of miming meanings when in communicative breakdown and of teaching the body language of the target cultures. Indeed, O'Dowd's (2000) earlier research concurred with Buckett et al.'s: for example, he describes an exchange in which students derived insights about the target cultures from observing the body language of their remote counterparts. On the other hand, Zähner, Fauverge and Wong (2000), writing in the same timeframe as O'Dowd, confirm Goodfellow et al.'s view that delayed transmission has a disruptive effect on turn-taking, and that body language and facial expressions are less useful to those watching the screen than hoped-for, because of various factors inhibiting 'natural' conversation.

In his literature review O'Dowd also reports a consensus around Goodfellow et al.'s advice that the videoconferencing sessions should be integrated within a wider pedagogical frame (Kinginger, Gourves-Hayward and Simpson, 1999; O'Dowd, 2000; Zähner, Fauverge and Wong, 2000), that is, they needed to be preceded and followed by related activities offline.

Normalisation and the improved sophistication of videoconferencing technology are factors that need to be taken into account when discussing the way that the feedback loop functioned over the years in the area of videoconferencing. We return to these issues in the conclusion of this chapter. However, other issues of relevance to the practice-and-research cycle emerge from a detailed comparison of the 1996 and 2006a and 2006b projects.

12.2 Goodfellow, Jefferys, Miles and Shirra 1996

12.2.1 Which research frameworks informed the study?

As part of the rationale for the use of the medium, the 1996 study cites arguments which indicate (in our interpretation) that the experimenters were guided by a communicative pedagogy with a language-acquisitional focus: the chosen teaching model is one where synchronous feedback is valued as liable to maximise acquisition, and it is also one where qualities of 'naturalness' and a broadening of the range of language skills through visual contact are valued, since they are seen as reflecting the situational authenticity that communicative language teaching prizes. Additionally, the researchers used collaborative learning theory, particularly in its relationship to the videoconferencing, which at the time of writing had been exclusively associated with transmissive rather than collaborative pedagogical activities. Thus the authors asked the following research question:

> [As] the pedagogical considerations inherent in managing collaborative work are different from those involved in organizing and delivering a lecture, remotely or otherwise ... the question arises whether remote collaborative working has more in common with face-to-face collaborative working, or with remote lecturing? (1996: 7)

In other words, this study was working within a third framework of investigation: the educational affordances of technology.

12.2.2   The setting

The study reports distance learning of English for Professional Purposes (for the Norwegian insurance industry) at a small private language school. Two locations were linked by videoconferencing. At the London site were a bilingual English–Norwegian language teacher and an English insurance expert, while another bilingual language teacher and six Norwegian students with lower to upper-intermediate English proficiency were at the Oslo site. Two principal activities were offered via this video-link. The first was a teacher-led class exploiting grammar and vocabulary as part of a traditional communicative approach. The second was an expert interview carried out by one of the teachers and designed so that students could put questions to the expert. The videoconferencing sessions were an experimental part of a larger blended course which was paper- and telephone-based, with the video-based component following on from the paper and telephone component in order to create further language opportunities, particularly through the input of the guest expert. Other materials (cassettes, worksheets) were used pre-sessionally with the intention of preparing the learners for the video experience.

12.2.3   Insights from practical application

In their conclusion the authors tell us that the intended communicative outcome was achieved, at least in terms of the ability of learners to participate in question-and-answer sessions in the medium. Very little peer discussion (i.e. two or more students alternating) was achieved. It is not clear whether the authors attribute this to the technology or to the pedagogical model used. During the sessions the researchers observed few instances of the student techniques practised in pre-sessional preparation, but a peer discussion did occur in the final debriefing, when the Norwegian students were encouraged to interact in their L1 with the London-based participants who were speaking in English. The authors conclude: 'this part of the session demonstrated that a non-lecture interaction was possible, and poses the pedagogical challenge of making it happen entirely in L2 rather than bi-lingually' (1996: 11). The staff involved included two tutors and one expert for only six students; much time was spent by students as well as staff, and many materials were produced for use as stimuli for the videoconferencing sessions, a time investment which needs to be set against results that proved at best modest. A major part of the conclusion relates to the large degree of interference from the technology that was observed. Here are two examples with consequences on teaching and learning: in the first a participant nods intermittently when others are speaking, and another later observes: 'Since the synch is slightly out we are unsure what he is nodding to' (1996: 12). The authors conclude: 'What is made explicit here is the difficulty of interpreting some of the unconscious behaviour of participants in the conference. This behaviour would not be evident in the absence of the visual dimension' (1996: 12). The second example concerns the frustration of the London-based tutor

> at not being able to signal that he was opening a question out to the wider group, rather than continuing in a one-to-one conversation. In a face-to-face situation this is done simply by moving the gaze around the room. In the conference, if the camera is pointing at someone, they end up having to answer the question. (1996: 12)

It is clear that the inflexibility of the technology had discursive consequences.

12.2.4 Feedback loop to research and further practice

From this study three main lines of reflection can be drawn regarding (1) progress towards language acquisition goals, (2) the way that technology may inhibit discourse mechanisms, and (3) technology and the distortion of body language as perceived by viewers:

1 To ensure progress towards acquisitional goals, integration of technology into the course is essential: a 'videoconference should ... build on previous teaching and learning and lay the ground for subsequent work' (1996: 15).
2 Regarding technology's role in inhibiting normal discourse mechanisms, the authors' advice is 'that the contributions of learners, especially the less confident, should be more explicitly structured into the interaction ... [and] that reflective teaching and learning activity, such as correction etc. may best be reserved for "after-the-event", for example watching the recording for self-evaluation, going through the tutorial again and picking out relevant points' (1996: 14).
3 On technology and distortion of body language, the study shows 'that language learners may have to be in some way prepared for a videoconferencing session, taught to use verbal rather than visual cues to exchange turns, and perhaps given a set of guidelines on what kind of body language to use, and how to dress, even, so as to maintain the highest visual quality possible and the smoothest flow in interaction ... [and] that those who manage

camera viewpoints will have to develop skills in representing the dynamics of a group interaction, according to whether one-to-one or one-to-many interactions are predominant' (1996: 15).

Overall, then, the 1996 study showed that videoconferencing can amount to more than a remote lecture, but that 'the language interaction it supports is in many ways different from the "face-to-face" equivalent' (1996: 16) and that mediation via this technology (at least in the state of advancement that characterised it in 1996) frequently distorted normal communication.

12.3   O'Dowd 2006a and 2006b

12.3.1   Which research frameworks informed the study?

O'Dowd uses the medium of videoconferencing in an investigation targeted at specific learning outcomes: the development of intercultural skills and of academic skills (i.e. ethnographic methods). One of O'Dowd's main foci is the emerging area of intercultural CMCL (see section 1.5). His research question is whether 'videoconferencing can make a particular contribution to intercultural telecollaboration that other communication tools such as e-mail or chat can not' (2006a, 94). In other words, what are the educational affordances of the medium for intercultural learning?

His second focus is on ethnographic methodology as a learning outcome. In this respect he makes the point that asynchronicity has been claimed to be of great use in cultural investigation because there is time available for ethnographic interviewing and for support to the trainee ethnographer. This consideration leads him to ask whether student ethnography is an appropriate method for videoconference-based interviewing, especially given the probing and sometimes intimate nature of the questions needing to be asked. In other words, what are the educational affordances of the medium for student ethnography?

Finally, unlike Goodfellow et al., who were working without the benefit of prior research, O'Dowd was able to use earlier publications (e.g. Kinginger, 1998; Butler and Fawkes, 1999; Zähner, Fauverge and Wong, 2000) to discuss the relationship between different configurations of videoconferencing (e.g. one-to-one, one-to-many, many-to-many) and particular learning outcomes, such as tandem learning, international telecollaborative projects or heritage learning (about the latter three categories, see also Thorne, 2006: 7–8).

12.3.2 The setting

Reported in studies in 2006a and 2006b, O'Dowd's project involved 25 German-speaking students of intermediate-advanced English at a German university, working with 21 US-based non-Germanists reading Communication Studies at a US university. Two tutors worked at each site. The partnership included the video sessions plus an email exchange of prescribed volume and content (though the emailing was not concurrent with the video session). As with Goodfellow et al., other instruments (e.g. pre-session questionnaires) were used to disseminate information to both cohorts. The task was broken down into information-gathering about national cultures and about ethnography, followed by four group videoconferencing sessions over eight weeks, and finally individual reporting on topics relevant to the experience. The videoconferences were designed as group discussions, while the email exchanges (one-to-one) were meant to be ethnographic interviews, with interviewer and interviewee alternating at some point.

12.3.3 Insights from practical application

The visual qualities of the medium received a mixed response: some students experienced videoconferencing as if it was 'normal' interaction, 'closer to reality' (2006b: 198), enabling them to 'see on [peers'] faces what they're really thinking'. Some even felt that turn-taking was more efficient in this medium; but others could not recognise facial expressions on their screens (2006b: 198) while yet others experienced the immediacy as oppressive ('I sometimes felt like in court', 2006b: 111). For those admiring the lifelike nature of the medium, it helped them to bond with their partners and taught them facts about their correspondents' cultures through observation of their body language.

The immediacy of the response demanded by videoconferencing provoked strong emotion and tensions. Some staff and students felt that this should have been avoided, and could have been achieved by allowing participants more time to monitor and mediate their feelings. Yet the emotional and tense moments were thought by other students to have been educationally beneficial as they were indicative of the strength of societal values held by their partners. Overall, though, it is not clear whether the study's insights into socio-affective rapport and management of ambiguity relate to videoconferencing specifically rather than to the time pressures associated with synchronicity in general.

A related insight came from the observation of students performing their ethnographic interviewing tasks. O'Dowd found that many were unable to adhere consistently to the non-judgemental ethos expected in ethnographic research. He speculates that it became obvious that students abandoned objectivity during the video sessions because of the medium's role in making it impossible to avoid or ignore awkward feelings: 'They were, in a way, obliged *by the nature of the medium* to delve further into the topics in hand' (2006b: 203; emphasis added).

Regarding the integration of email and videoconferencing, students held positive views. They felt that the two media were complementary, since partners were more relaxed through having seen each other and were able to discuss orally ambiguities that had troubled them in the email exchanges. For tutors, the combination allowed for complementarity between fluency (quick-fire turn-taking) that might sometimes produce 'superficial' cultural insights (2006a, 105) and cultural thoughtfulness (evidence supporting this hypothesis is that greater reflective quality was found in the emails).

12.3.4   Feedback loop to research and further practice

Like Goodfellow et al.'s, O'Dowd's first recommendation is for greater integration. Group-to-group videoconferencing should be supported by another medium, such as email. O'Dowd does not add – but we nevertheless suggest – that other communication systems might be used in future projects, providing there is complementarity between them (2006b: 198).

Responding to the issues raised by the 'nature of the medium', such as emotionality and the stresses of immediacy, O'Dowd advocates better learner preparation, in terms of methodological training (in ethnography) and of media training. Also, building on his findings about student failure to remain non-judgemental in their interviewing, he draws from the experience of various guiding principles for future ethnographic task design, a contribution to good practice that appears to be independent of the learning medium.

12.4   Conclusion

We suggest that some normalisation of videoconferencing occurred between 1996 and 2006. Bax describes seven stages of normalisation of technology, from 'early adopters' to full normalisation, with a penultimate stage, when the technology is 'normalising' (2003: 24–5). We speculate that the 'early adopting' team of 1996 was under implicit pressure to evaluate videoconferenced teaching in relation to face-to-face lecturing, whereas O'Dowd was working in a period when the technology was normalising and so came to medium-independent conclusions, leaving him free to concentrate on content-related skills – in this case, ethnographic methods.

Yet in spite of the historical difference between them, the two studies come to at least three identical conclusions: (1) that close integration is needed between the videoconferencing sessions and the other components of the course, on- or offline; (2) that students have mixed feelings about seeing themselves and their partners on screen; and (3) that much more training is required. There is some justification in being concerned that such issues appear to be unresolved after a decade of research. Judging by the quality of the literature reviews in the later works, this weakness in the feedback loop is not due to poor dissemination. Instead, we suggest that the unresolved issues endure because they have not been sufficiently problematised and that perspectives from cognate fields may be required in order to help theories develop. For example, insights from social psychology, communication theory, applied linguistics or multimedia semiotics might help to probe the reasons behind participants' diverse responses to technology-mediated images of self and others. Input from these disciplines may also help to structure our thinking about training, moving away from purely technical concerns and asking: What is the exact nature of the training required in order to learn to make sense of individual and collective discourses in a technology-mediated situation?

To conclude, our comparison of two projects supported by high-end technology shows that researchers have to different extents raised – though not resolved – issues that belong to the specific affordances of videoconferencing, but that the main concerns for us as reviewers of this literature remain the uncertain functioning and the slow pace of the feedback loop.

13
Emerging Technologies

As we mentioned in the Introduction to Part II, this chapter is different from chapters 8–12: as the media involved have only recently emerged, neither pedagogical implementations nor research studies are yet numerous enough to allow us to discuss the feedback loop in regard to these media. In what follows we therefore simply outline the emerging technologies' potential for practice and for research.

13.1 Blogs

Alongside wikis (section 13.2) and mobile learning devices (section 13.3) blogs – also called weblogs – are among the recent additions to the language teacher's and learner's toolbox. Blogs are a type of website that allow for the publication of text, images and sound files. Blood defined a blog as 'a website that is up-dated frequently, with new material posted at the top of the page' (2002: 12). While today's blogs have many different designs, their enduring distinctive characteristic is that they are designed to be easily updated, with the latest changes clearly visible to visitors. Blogs started as publicly accessible personal journals for individuals (*Webopedia* http://www.webopedia.com/TERM/b/blog.html), yet today they are also used for collective writing. While the purpose of traditional journals tends to be private self-reflection, blogs have become global communication tools. Their success can be traced to the following: several companies provide free software; blogs are easy to create; the Internet allows for global dissemination; and the 'comment' functionality makes them a two-way communication tool. So blogs have become a major phenomenon, used by millions to publish their online diary; by celebrities to promote themselves; by people to talk about their particular interests; by journalists to give an alternative view to mainstream, institution-based news; or by the mainstream media themselves to create (a semblance of) access for their audience. At the beginning of 2007, over 60 million blogs were identified by the blog search engine *Technorati*.

In the educational context, blogs are still relatively untried. However, teachers increasingly see their potential for creating a collaborative learning environment, providing learners with a language forum to exchange with peers and to reflect on their work, and to foster learner autonomy and learning strategies (Batardière and Jeanneau, 2006: np). As yet, there are few research studies on the use of blogs for language learning. Many such papers are descriptive; they deal with the characteristics of the tools and sometimes point out the potential for learning. Campbell (2003), for example, suggests that blogs are useful in an ESL context as a tutor blog, a learner blog or a class blog. Taking up Campbell's suggestion of learner blogs and his recommendation about their suitability in reading and writing classes, Wu (2005) introduced blogs to enhance an EFL writing class in Taiwan. Wu points to certain functionalities of blogs, namely 'easy-to-use interface, frequent text update, and interactive comment area' (2005: 1). Findings of the small survey that Wu conducted with the students suggest that whether blogging is effective in language learning and teaching depends to a large extent on how it is used in a course.

In the introduction to his paper on blog-assisted language learning, Ward focuses on the 'completely new form with un-chartered creative potential' (2004: 3), using the concept of 'voice' to distinguish it from the more conventional journal. '[T]he weblog's ability to accommodate multiple authors provides more dimensions and generates a different kind of discourse than the traditional journal' (2004: 3). Ward's study of using blogs in ESL writing classes, however, only addresses the area of student perception of blogging. We believe that the concept of voice is a promising avenue for examining the functionalities of the medium and exploring what it affords the language learner.

Researchers have also started to consider the use of the specific functionalities and affordances of blogs from a more theoretical point of view, in the main inspired by sociocultural thought. Pinkman (2005) uses the concepts of learner empowerment, learner autonomy and learner independence as a starting-point for her study, emphasising the importance of learners taking responsibility for their learning if they are to become effective. Although Pinkman set out to show how blogs can be used to encourage learner independence, her project was unable to do so (for a description of her procedure, see section 14.2.4 and

Figure 14.2). One the one hand, this was due to the small number of participants; on the other, the study shows that it is not sufficient to provide learners with an online tool such as a blog and expect them to develop learner independence. Pinkman suggests that its introduction needs to be carefully planned for and integrated with other curricular activities; students need to be prepared and supported; and both teachers and learners need to be aware of the possibilities and the limitations of the tool – for example, the fact that blogs are usually open for anybody to read and comment on. As the author indicates, cultural practices may play a role and 'more work needs to be done on assessing the attitudes of university students in Japan toward learner independence and out-of-class learning, to determine the best methods to use in order to encourage learner independence' (2005: 19).

Murray and Hourigan (2008) examine two possible approaches to blogging, namely the socio-cognitivist and the 'expressivist'. The former focuses on collaborative knowledge construction, while the latter emphasises the creative and reflective benefits of blogging. The study shows that the socio-cognitivist approach is beneficial for group projects and for building confidence, motivation and learning; the 'expressivist' approach is useful for giving learners, particularly more advanced learners, more freedom and control.

Though not specifically related to language learning, Lankshear and Knobel examined blogs in the context of new literacies. In Lankshear and Knobel (2003b) they discuss two approaches. The first emphasises the potential of blogs to promote 'powerful writing' in the context of genre theory and critical literacy; the second suggests the possibilities of blogs as 'indices to and evidence of personal and collective knowledge structures' (e.g. in a school context). More recently, Lankshear and Knobel have looked at the social practice of blogging as an instantiation of 'the deeply participatory nature of these [new] literacies' (2006: np).

13.2   Wikis

While wiki technology is becoming increasingly popular both inside and outside educational contexts, there are few reports of studies on the use of wikis in language learning. So what is a wiki and what distinguishes it from a blog? The *Webopedia* (http://www.webopedia.com/TERM/b/blog.html.) defines it as a

> collaborative Web site [which] comprises the perpetual collective work of many authors. Similar to a blog in structure and logic, a wiki

allows anyone to edit, delete or modify content that has been placed on the Web site using a browser interface, including the work of previous authors. In contrast, a blog, typically authored by an individual, does not allow visitors to change the original posted material, only add comments to the original content.

The main functionality of joint authoring is exemplified in the *Wikipedia* (http://en.wikipedia.org/wiki/Main_Page) the free online encyclopedia, which in 2006 featured six million articles in many languages. The example of *Wikipedia* shows that a wiki may be a tool with potential for collaborative and problem-based learning, two major concepts in sociocultural theory.

Engstrom and Jewett (2005) investigated collaborative learning in the context of a geography class and their findings are applicable for learning more generally. Their study was informed by notions of inquiry-based learning where students work collaboratively on a problem. Wikis seem a suitable tool to realise this kind of learning because they constitute 'collaborative environments by design, and can serve a variety of purposes for collaborative online projects' (2005: 12). Learners are able to create and edit content collaboratively, and most wikis have an edit trail, so that students as well as teachers have access to every version of the website. Engstrom and Jewett group the lessons learned into instructional issues and technology-related issues. For example, the wiki pages reflected mainly surface-level thinking, which the authors explain by proposing that teachers may not have been modelling or facilitating 'an exchange of ideas, questions and feedback across school teams on the wiki pages' (2005: 14). Engstrom and Jewett conclude that 'the teachers could benefit from more practice in prompting student's critical thinking through the use of information literacy skills' (2005: 14–15). The technology-related issues revolved around problems with access to the wiki and the restricted use of computer labs. This study shows the potential of new technologies for collaborative and inquiry-based learning. At the same time, it highlights several issues which go beyond the context in which the study was set, issues which apply equally to language learning settings:

1 the importance of the teacher's role in supporting students and in facilitating the critical inquiry process that was at the centre of this project;
2 the need for training; and
3 the need for technical support for teachers as well as students.

A study by Lund and Smørdal (2006) focuses on collective cognition – 'the process we engage in when we collectively develop insights' – and asks 'to what extent a wiki can be conducive to collective knowledge building and, in particular, if and how a teacher can take part in wiki activities' (2006: 37). Lund and Smørdal's is also a sociocultural approach, based on concepts such as zone of proximal development, activity theory, community of practice and collectivity of practice. These concepts informed an intervention study at a Norwegian school, which aimed to foster learning through the use of digital and networked technologies – in this instance a wiki was used to support collective learning. The study examined learner production, individual mastery of the application, strategies used and the role of the teacher. The authors sum up the challenges as follows:

> Working with wikis involves an epistemological shift, from individually acquired to collectively created knowledge. ... It follows that the teacher's professional repertoire is expanded. Planning lessons, a traditional hallmark of teacher expertise, need to be extended to *designs*. (2006: 44)

13.3   Mobile devices

Although language learning with mobile devices was documented eight years ago (Godwin-Jones, 1999; Brown, 2001), it is only recently that studies have began to appear in a field that Chinnery (2006) and others have called MALL (Mobile-Assisted Language Learning), in a punning adaptation of Warschauer's (1999b) remark on acronyms. Why, Warschauer had asked, do we need to give the *computer* the all-important role reflected by the acronym CALL, given that we don't single out other mediational tools in this manner? For example '[w]e have no "BALL" (book-assisted language learning), no "PALL" (pen-assisted language learning), and no "LALL" (library-assisted language learning') (1999: np; original emphasis). Today, the question is just as relevant: why single out mobile-assisted language learning as a practice distinct, say, from sedentary language learning, which you do at your desk, or ambulant language learning, which you might do by walking around with L2 native speakers in the streets of an L2 city? And what do we mean by mobile-assisted language learning? In the following paragraphs we start by providing a definition of what we prefer to call 'mobile device-assisted language learning', before discussing the reasons why we think that it should be recognized as a field of enquiry.

In her definition of mobile devices, Kukulska-Hulme (2007) includes 'cellphones, personal media players, personal digital assistants (PDAs), smartphones and wireless laptops' (2007: 119). We explore these tools with the exception of the laptop. A laptop is certainly a movable object (and capable of spending some time in wireless use), but many of its affordances are similar to those of the computer, since its screen, keyboard and sometimes mouse or webcam attachment are close in size and design to those of desktop computers. Therefore much of what we, and the researchers cited in chapters 8–12, have said about projects carried out with desktop-based tools is applicable to laptops.

Specific (potential) affordances of mobile devices for language learning include what Kukulska-Hulme calls 'contextual learning' (2007: 123), a type of learning exemplified by schoolchildren collaboratively producing guided tours of locations while on the move in those locations, as 'a handheld device with GPS capabilities delivers location-sensitive information when a child walks into node areas indicated on a map' (2007: 123). The applicability of this type of educational scenario to a visit to an L2-related location is obvious, but as yet under-reported in the literature.

Another way of understanding the specificity of mobile device-based learning is to think of the ubiquitousness of many of the devices, which allows them to be used for at least two types of situations, illuminatingly captured by Colpaert: 'Through online or mobile services an individual can communicate with people who are aware that he/she is learning a language, which – automatically or not – entails some kind of mediation. The second type of interaction is communication with people who are not aware that the individual is learning a language and who therefore provide no explicit mediation' (2004: 263). Some analytical literature is available regarding the first, mediated, type of interaction (e.g. Kiernan and Aizawa, 2004, on problem-based-learning on mobile phones; Lan, Sung and Chang, 2006, on collaborative reading using Tablet PCs). However no literature has yet been forthcoming to document Colpaert's second type of situation, although such learning settings could be of great interest not only because they are ecologically valid but also because they create a situation to which mobile technologies are particularly well adapted.

In their summary of a survey recently commissioned by the UK-funded Joint Information Systems Committee, Kukulska-Hulme, Evans

Table 13.1 Mobile-devices for languages: practitioner priorities

* Consider a repertoire of possibilities for the new technology: its potential to support teaching, learning, and the management of teaching and learning.
* Review how wireless and mobile technologies might facilitate contextual learning in your subject, i.e. allowing the information available in a learners' location, and relevant to their needs, to be captured or delivered in context and to contribute to teaching and learning.
* Investigate the scope for continuity of learning, i.e. taking advantage of availability of a portable device in an institutional setting, workplace setting and at home, where this can encourage consolidation and increased familiarity with learning material.
* Appraise the various communication channels between yourself and your students, e.g. SMS, voice messages, email, online or mobile discussion forum, from a social as well as a pedagogical point of view.
* Be cautious about claims that the new technologies can be used 'anywhere, anytime': pedagogical, technical, logistical, usability, and social constraints must not be overlooked.
* Consider the physical environments in which new technologies will be used, and how this could impact on effective learning.
* Make time to understand new student audiences and patterns of study that emerge when learners obtain access to wireless and mobile technologies, including non-traditional entrants.
* Exploit the support that mobile devices offer to social networks, communication with mentors and experts, and interaction in online communities.
* Explore how mobile and wireless and mobile learning can make for a more immersive experience in your discipline, through increased richness and diversity of both content and activity.
* Remain on the lookout for unexpected benefits or learning outcomes, as well as unanticipated disadvantages.

(Kukulska-Hulme, Evans and Traxler, 2005)

and Traxler (2005) identify the priorities for teachers intending to make use of wireless and mobile technologies in Table 13.1:

With a few exceptions, in the literature on mobile device-assisted language learning in the area that is core to this book, interaction has moved from describing usage (Dias, 2002) and resources (Godwin-Jones, 2005) to analysing benefits and shortcomings through empirical learner data. To approach such data in both types of situation imagined by Colpaert, researchers might do worse than to draw inspiration from Kukulska-Hulme et al.'s (2005: np) recommendations to practitioners.

13.4   Conclusion

This chapter has illustrated the possibilities of emerging technologies for collaborative, autonomous and critical language learning and gives examples of the educational approaches that may be applied. However, researchers have only just started exploring how to realise this potential and what the challenges are. Studies such as Pinkman's (2005) show that online tools such as blogs do not automatically motivate students and turn them into independent learners who use the L2 outside the classroom; institutional as well as cultural factors also play a role. We have also seen that time, motivation and training are needed on the part of teachers (Engstrom and Jewett, 2005), and that task design and intercultural issues play a role as well as the way such tools are integrated into a language course. There is scope for a great deal more research, empirical as well as theoretical investigations.

Part III
Practitioner Research

14
An Overview of Practitioner Research

14.1 What is practitioner research?

At a roundtable discussion on the last day of the 2006 EuroCALL Conference, Diana Laurillard expressed the view that the best chance the CALL/CMCL community had of influencing the future was to ensure that it took every opportunity to clarify and know what learners need, acquiring and disseminating this knowledge. This, according to Laurillard, the CALL/CMCL community could achieve through the sheer strength of its practitioner research potential. We agree, as did many in the conference audience. This section defines practitioner research, drawing out its two main strands: action research and exploratory practice.

A wealth of valuable methodological knowledge about practitioner research is available, born of the experience of the offline community of language professionals, and there is no reason why the advice generated in that field should not be applicable to CMCL. This is why this chapter, in which we outline two types of small-scale research and briefly describe research and data collection instruments, is written with both off- and online practitioners in mind. In chapter 15 we will be concerned to identify the implications for the small-scale practitioner researcher of the fact that the researched participants are online.

14.1.1 Action research

Action research, a methodology originally arising from social psychology (Lewin, 1948) which was later extended to many domains of the social sciences including education, is described in many research textbooks, some of which you will find listed in the Further Reading section of this chapter, and in websites (see Part IV). One definition that we have found particularly clear is Benson's (2001: 182), which we reproduce here in its original boxed format almost *in extenso*.

Quote 14.1   Action research according to Benson

Action research has five distinctive characteristics:

1 It addresses issues of practical concern to the researchers and the community of which they are members.
2 It involves systematic collection of data and reflection on practice.
3 It is usually small-scale and often involves observation of the effects of a change in practice.
4 It often involves analysis of qualitative data and description of events and processes.
5 Its outcomes include solutions to problems, professional development and the development of personal or local theories related to practice.

In language education, the action researcher is often a teacher acting in the role of teacher-researcher. In collaborative action research, teachers work together on shared problems. Burns (1999: 12) states that the goal of collaborative action research is 'to bring about change in social situations as a result of group problem solving and collaboration'. She argues that collaboration increases the likelihood that the results of research will lead to a change in institutional practices ...

(Benson, 2001: 182)

14.1.2   Exploratory practice

'Exploratory practice' is the term used by Allwright (2005) for what he calls an epistemological and ethical version of practitioner research, which he felt compelled to set up in opposition to action research, when his leadership of a practitioner research project in Brazil proved less fruitful than he had hoped. He explains why.

Quote 14.2   Disadvantages of research for teachers according to Allwright

[The] research project was clearly taking up far too much staff time to be worth pursuing, and it was also requiring staff to learn research skills that were not likely to be helpful in their lives as teachers. So it was heavily parasitic upon their normal working lives, rather than supportive of them, or integrated into them. To make matters worse, my weekly workshop on classroom research skills (a highly technicist enterprise that put how to do research above all else) was spreading this academic view of research and asking teachers, outside the official research project, to add a time commitment and the obligation of learning academic research skills to their already extremely busy professional lives.

(Allwright, 2005: 357)

Exploratory practice (EP), according to Allwright, puts understanding a situation above solving a problem. This sets it apart from action research, which is resolutely oriented to solving and improving. Another important principle of EP is that it uses classroom activities rather than academic research techniques, and that learners are practitioners as well as teachers. Finally, EP is more concerned with the quality of a learner's or teacher's life than with the quality of output, on the principle that if you improve the quality of life, you are more likely to create the conditions for performance improvement anyway.

In the rest of this section, we put some flesh on the bones of these definitions by setting out the three stages of good practitioner research, then show, in diagram form, how the three-stage outline has been applied by three practitioner-researchers. The first two, Clerehugh (2002) and Pinkman (2005) call their projects 'action research', and the third, Zhang (2004), 'exploratory practice'. While none of the studies is in all particulars a canonical example of the genre – indeed, there is no such thing – we hold them up as simple models for a variety of small-scale, local projects.

First we turn to the three steps that characterise practitioner research, starting with the crucial proviso that, rather than follow the three-step design linearly, projects frequently go through the stages cyclically, in a 'spiral of steps' (Burns, 2005: 58), with the aim of testing successive improvements to any one situation, iteratively and self-reflectively.

14.2   The three essential steps of all practitioner research

The three steps provide the structure of a resource that we have used in what follows. This resource, the 'Embedding Learning Technologies' website, aimed to encourage higher education teachers to embed ICT in their teaching. The site has not been updated since 2003, so some of its links are now inactive. However, it still constitutes a rich resource that is applicable to CMCL practitioner research. The main address is www.elt.ac.uk/materials.htm, but see our notes on its still active links in Part IV.

Step 1: reviewing your practice, or 'What do you want to find out and why?'

Although this may appear to be a dauntingly vast question, many researchers do not start with a blank sheet and an unbounded agenda of discovery. On the contrary, they may be constrained in their choices by a variety of factors. These in themselves can be useful in shaping your thinking from the outset. They may include:

- identifying improvements which could be made to a specific course, learning activity or learning resource;
- reflecting on professional practice in a structured way;
- building evidence for a portfolio (e.g. career development, teaching fellowship);
- producing guidelines for colleagues (internal and external) who might want to carry out a similar innovation;
- generating data for a research study or publication, perhaps as part of your professional development studies;
- investigating an issue of personal, intellectual or professional interest;
- satisfying internal or external auditing requirements.

Also useful in orienting the direction that your project will take is the question of who has an interest in your findings: students, fellow teachers and those with a professional stake in learning technology, technical staff, present or future employers, managers, advisers or inspectors. The framing of questions will vary depending on which of these stakeholders you wish to inform or influence. Additionally, as the Embedding Learning Technologies website points out,

> any of these people can provide data to help you to evaluate your project. They can also act as co-evaluators, for example by:
> - helping you define your evaluation objectives and goals
> - observing your teaching with technology
> - collecting data on your behalf
> - road-testing questionnaires, structured interviews and other data collection instruments
> - acting as witnesses, mentors or critical colleagues – particularly useful if you are using an action research approach.

As well as these predeterminations, you will have in mind concerns that arise out of your everyday teaching. Below are some indicative examples, which it may be helpful to organise into categories as in the five headings below. You will undoubtedly have other categories and concerns, which you might like to add for your own project preparation.

1 Classroom management

- Did the resource that I used help me with the large/small group size that I have to manage?
- Am I achieving an appropriate result in terms of group dynamics when teaching via CMCL?
- ...

2 Materials and tasks

- Is there evidence that the CMCL resource and task that I used supported a range of learning styles and needs?
- How could the design of this session/course/resource be improved?
- What supplementary materials will motivate my students to work in-between online sessions?
- ...

3 Particular skill or area of knowledge

- Is there evidence of my students developing new skills? What are they?
- Did the use of CMCL help my students acquire transferable skills?
- ...

4 Student behaviour, achievement or motivation

- What are my students' attitudes to communicating via computers?
- How well do my students learn when using CMCL?
- How enjoyable was the experience of communicating online for my students?
- How can I build my students' image of themselves as online communicators?
- ...

5 Personal professional issues (e.g. time management, relationships with colleagues/managers)

- How much time did I invest in preparation/support/follow-up?
- What new skills did I need?
- How effectively was I trained/did I train myself to carry out the CMCL teaching?
- Am I satisfied with the way I evaluate my students' online communication work?
- ...

Having devoted some time to thinking through these issues you may be ready to formulate a precise research question and to structure your project's outlines by using a checklist such as the one below from the Embedding Learning Technologies website, where further details can be obtained on each of the steps in the checklist (see Part IV).

Checklist 14.1   Outlining your practitioner project

1   Formulating the question
Why am I carrying out this evaluation? (identify objectives; designate stakeholders)
What information do I need? (define evaluation goals or research questions)

2   Collecting the data
Who can provide the information? (identify resources, e.g. students, other staff involved in the project, an independent observer)
How can I best collect this information? (choose a data collection strategy)
When should it be collected? (e.g. before, during and after a learning activity, or at different points in a module)

3   Analysing the data
How will I analyse the data? (select appropriate analysis techniques, bearing in mind the nature of the data and the evaluation goals)

4   Drawing conclusions and reporting on findings
What does my evaluation mean? (reflect on implications for own practice; draw recommendations or lessons for others)
Who needs this information? (which of the stakeholder categories; consider also the wider disciplinary, institutional and professional communities)
How can I reach them? (identify dissemination opportunities)

(ELT Embedding Learning Technologies website)

Step 2: design and implementation, or 'How will you carry out the actions and how can you pre-empt difficulties?'

Putting on paper or screen a flowchart of events and milestones planned to occur as your project develops is a good way of anticipating possible pitfalls. This will be more effective if you submit your flowchart for comment to a critical friend, colleague, mentor or student and modify it according to their response(s). Items on the chart will include:

- a full description of the action or innovation to be implemented;
- a timeline for implementation (including the development of instruments and the briefing of participants pre-implementation, and the follow-up post-implementation);

- the membership of the participants, including co-researchers, observers and technicians;
- an exhaustive list of resources (hardware, software, paperwork: invitations to participate, worksheets, questionnaires, ethical permissions forms, etc.) without which the plan cannot proceed;
- a description of the data to be collected (type, volume, medium, storage needs);
- a data collection and analysis plan (where, how, from whom and at which point);
- a description of the chosen instruments (see section 14.3) and tools (including backup servers).

Step 3: evaluate and disseminate, or 'What have you found out and who needs to know your conclusions?'

A clear illustration of the merits of simple and complex data analysis methods is offered by the Embedding Learning Technologies website:

Quote 14.3   Simple or in-depth evaluation of data?

Having collected the information, you need to ask what it means in the context of your original evaluation issue or research hypothesis. This might be quite straightforward. For example, in the case of [answers on numerical scales], it is a simple matter to find the mean of subject scores and to compare responses on different issues. You will be able to conclude, for example, that students rated the discussion board more highly than the links section in your online learning environment. You still need to be aware of how to interpret this finding. Why did they rate this aspect of the experience more highly? Do such ratings actually translate into effective learning? Triangulating with other data such as transcripts of online discussion, will help you to add depth to this finding.

(Embedding Learning Technologies website)

Triangulation brings together data from two or more data collection methods in order to illuminate the interrelationships involved in the overall, usually complex, educational picture of the situation you are researching. As you add depth to your project, you may wish to consider further methods of quantitative and qualitative analysis, possibly with the help of web-based and other tools (see chapter 16).

Finally, your project needs to reach a phase where you draw conclusions and present them to others. Here are some of the questions that you need to ask yourself before firming up your findings.

Quote 14.4   Preparing your conclusions for dissemination

- To what extent do the data validate the hypothesis or answer the question? Is a different hypothesis or question called for?
- Did my intervention make a difference to the student learning experience? What was the impact? How significant is it?
- Is it possible to assess which aspects of the learning situation were most effective?
- Are there significant differences among different classes of respondent (e.g. types of student, or between students and staff)? How significant are they?
- What patterns, if any, emerge from the data?
- Were any problems identified? Do any unexpected issues or findings emerge from the data?
- Are the data inconclusive or contradictory? How can I explain this?
- Is the quality of the data good enough (e.g. was there triangulation? are the data representative?)? What qualifications do I need to make when presenting my findings?
- What other data would help to explain these findings or to make the situation clearer?

Dissemination for much of the research referenced in this book means publication in order to reach a wide community of researchers. But it is always worth remembering that the term may have other interpretations and that using less conventionally academic methods may sometimes help your ideas circulate wider. For example, you may wish to make your findings known to your immediate colleagues via a talk, to your more distant professional community in the form of a conference paper or poster, to your managers through a report or to your students in a blog or presentation. Whatever method you choose, the Web offers guidelines and templates, some of which are listed in Part IV.

14.2.4   Three simple practitioner research projects in diagram form

The three projects in this section have a common starting point: the realisation by a teacher that her students were not learning, or not as well as hoped. In each case, the teacher saw computer-based networking as a possible answer.

The first study, by Clerehugh (2002), shows how a teacher of German tackled and improved underperformance by her secondary school pupils. How did she map out her 'small-scale action research'?

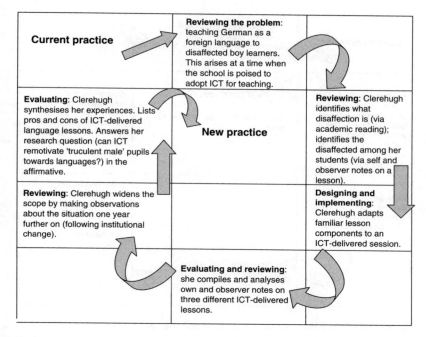

Figure 14.1   Small-scale action research to solve a motivation problem

Figure 14.1 shows the basic line that her action research took, starting with her perception of a problem in her teaching and some academic research by her on what she perceived as the root of the problem: school disaffection by male pupils.

The second study, by Pinkman (2005), explains how an EFL teacher working in Japan sought to enhance her students' educational experience. How did she understand 'small-scale action research'? Figure 14.2 (below) shows the basic components of Pinkman's project, which started not with solving a problem, but with the desire to explore the possibilities relating to the then novel tool family called blogs. Like Clerehugh, she grounds her action research in a theoretical dimension: learner autonomy.

How does exploratory practice differ from action research? Figure 14.3 (below) shows a project undertaken by Zhang (2004), whose starting point, like Clerehugh's, is a motivation problem. Unlike Clerehugh and Pinkman, however, he does not start from a theoretical *a priori*. Instead,

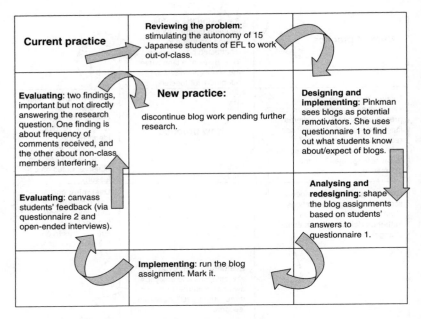

Figure 14.2   Small-scale action research to enhance autonomy

he relies on trial-and-error and his adoption of a principle of EP, according to which knowledge arises from the involvement of all concerned, allows him to understand his students' behaviour and to create the conditions for change.

14.3   Overview of methods and instruments

There is a very large volume of literature on educational research methods and instruments; we refer you to such works in this section and in section 14.5. The following is only a brief overview, providing an *aide-mémoire* to some of the more common methods and instruments of practitioner research, which can be used singly or in clusters depending on the scope of the project, the type of research question and the time available. Table 14.1 shows methods and processes that are likely to be relevant to you if you plan to set up a small-scale project, which may help you in the course of running such projects. These methods and

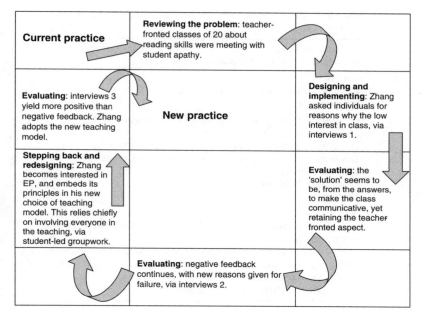

Figure 14.3   A small-scale exploratory practice project to gain understanding

processes are also listed in alphabetical order of the terms in bold, and briefly described.

Case studies 'are often fishing expeditions in which one hopes to clarify suspicions and disclose the unexpected. Case studies are more likely to make use of the collection and presentation of detailed information about a particular participant or small group rather than a large group representative of society as a whole. As such, a case study may not even have a clear research agenda other than attempting to document a subject, time and place as thoroughly as possible' (Beatty, 2003: 208).

Critical event recall, conducted in focus groups by using the researcher's records of the interactions that took place in class as a stimulus to the focus group members' memory, can stimulate participants to recall important learning events that they might otherwise have forgotten: 'When groups of participants engage in mutual or shared recall of events in which they have been present together, they can gain insight into their behaviour and learning processes' (Grabe and Stoller 2002: 167; De Laat, 2006: 56).

Table 14.1 Overview of methods and instruments for small-scale projects

Overall framework   **(Human) participant research**

Methods
    Ethnographic studies and **case studies**
    Experiments

Processes involving you (and perhaps other researchers)
    Carrying out a **literature review**
    Designing, distributing and collecting **questionnaires**
    Document gathering (including all paperwork related to your
    subjects' activities)
    Compiling **fieldnotes**
    Observation, peer observation

Processes involving you and your subjects
    Organising a **critical event recall** (or **simulated recall**) session
    Organising **focus groups**
    Carrying out **interviews** (of groups or individuals)
    Asking for **journals, diaries, logs** to be kept
    Asking for **narratives** to be written
    Carrying out a **think-aloud** session

Document gathering does not impact on classroom instruction; it simply involves 'gathering sets of documents that are relevant to the research question (e.g. lesson plans, pre- and post-tests, software, student exercises, worksheets' (Grabe and Stoller, 2002: 166). Additionally, session screen captures – still or moving – can be gathered (see also section 16.1).

Ethnographic studies are narratives in which the focus is on a social group. This group may be learners, teachers or a combination of the two. The purpose of an ethnographic study is to observe and understand their behaviour. Ethnographic principles can inform case studies.

Experiments are 'a way of finding out information by creating an artificial situation and changing the variables at play' (Beatty, 2003: 206). This is often done by testing something on one group (giving the group a particular treatment) and setting up a control group, which will be given no treatment, for comparison.

Fieldnotes are 'a written record of classroom events related to the research question … Fieldnotes are taken as the study proceeds, not after the fact' (Grabe and Stoller, 2002: 166).

Focus groups involve unstructured group interviews and are used to gather opinions about the value of procedures or products. The focus group is asked to reach a consensus on issues, and the choice of members (people with similar characteristics or a mixed group) is important. Group teleconferencing now makes it possible to hold such meetings online.

(Human) participant research is what educational action researchers are most likely to be involved in: research that uses human participants as subjects. However, participant research as a concept is much wider than both educational and action research since it includes a great deal of medical research. Ethical considerations are an essential part of the planning needed before undertaking such research (see section 15.2).

Interviews can be conducted face-to-face, or by telephone, email or video- or computer-conferencing. They are 'interactions conducted by the teacher in a structured, semi-structured, or unstructured format with teachers, administrators, librarians, aides and parents' (Grabe and Stoller, 2002: 167). Teacher–student interviews can also be conducted individually or in groups, providing an ethical framework has been set up (see section 15.2).

Journals, diaries or **logs** are a 'written record of teachers' opinions and reactions to research questions and related issues. Dated journal entries are usually completed after class' (Grabe and Stoller, 2002: 166). Blogs can easily be created and used for this purpose, particularly if the journal is to be read and discussed by more than one teacher. Journals, logs and reflective blogs can also be kept by students, at your request, and used with their permission in your research.

A **literature review** provides a background to understand the state of knowledge in the field and the context to which your question or the hypothesis at issue needs to relate. While this instrument is more suited to large-scale research, the examples of Clerehugh and Pinkman in Figures 14.1 and 14.2 show that a literature review can provide a grounding for a small action research project.

Narratives are accounts of experience that your participants give, in written or spoken form. To encourage them to reflect further on events, various methods can be used, such as working with metaphors, pictures or brainstorming. Think-aloud protocols and critical event recall methods can also contribute to a project narrative.

Observation is a way of studying a group at work. Observations may be made during the online event if it takes place in real time, but it is easier to work with recordings. If working in real time, the teacher will be very busy teaching, and so would be well advised to ask a peer to observe.

Peer observation involves your class being observed by a critical friend, who will have been well briefed by you on the aspects that you particularly want feedback on. The observations may be open-ended or standardised using a template you have designed. Group teleconferencing makes it possible for such observations to take place online.

Questionnaires appear to be easy to grasp as a research instrument as many of us have had the experience of responding to one. However, this is deceptive, as their content needs to be carefully thought out, and you have to consider whether all of your class(es) need to be involved rather than a sample, and what sort of question best serves your situation (Zemsky and Massey, 2004).

Simulated recall see critical event recall.

Think-aloud protocols are observations of learners engaged in an activity (e.g. using a computer program). Encouraging them to think aloud, to say what they are thinking and wondering about at each moment, and recording their comments, allows for reflectivity in action, as learners may be struggling with aspects of the interface and able to verbalise their progress.

14.4   Summary

This chapter is addressed to those interested in carrying out research with their own students; in other words, practitioner research. We have defined different types of such research and have identified three broad organising principles that any practitioner research project can be expected to follow: reviewing practice, designing and implementing action, evaluating (and possibly disseminating) findings. We have shown, via case studies, how these principles have been used by three specific practitioners to underpin their work. Finally, we have provided a brief overview of methods and instruments, which can be complemented by reference to the next section as well as to Part IV.

Further reading

From fields other than CMCL

Burns (2005). In this state-of-the-art article the antecedents, definitions, processes and purposes of action research in the field of English language teaching are discussed. Action research is also considered in relation to more established notions of basic and applied research.

Carr and Kemmis (1986). A foundational work of critical literacy, making robust connections between action research and social change, promoting the principle that educational practitioners have to be committed to self-critical reflection on their educational aims and values.

Dörnyei (2003). A practical book offering guidance to practitioners and researchers for compiling, administering and analysing questionnaires specific to L2 research.

Gass and Mackey (2000). Gass and Mackey explain the use of event recall methods in the specific context of second language research.

Kemmis and McTaggart (1988). An influential book setting out the basic model for action research: plan, act, observe, reflect; after which, plan again for the next iteration.

McDonough (2006). A nice example of a small-scale study using language graduate teachers' professional journals, reflective essays, action research reports and interview feedback. A reflection on integrating guidance on action research into graduate programmes is offered.

Silverman (1997, 2001). These companion volumes are classics of methodology writing. A sociologist who focuses on research from face-to-face settings, Silverman contributes much information which can be adapted to small project-based CMCL research.

15
A Practical Guide to CMCL
Practitioner Research

In this chapter we concentrate on the practicalities of CMCL research to offer support to readers interested in setting up projects in this field. From one point of view, conducting CMCL research is like any other education research experience: human participants (students and teachers) have to be selected in a principled manner and treated ethically; data protection concerns have to be met; data must be collected and selected in ways appropriate to the pedagogic setting and to the research objectives. However, in CMCL, research is conditioned by the computer-mediated nature of the experience, which affects the participants, the conditions in which they are networking, the data they are producing and the practical aspects of data collection and analysis.

While the great variety of practicalities involved can be a source of complexity that may make the prospect of setting up and running a project seem daunting, it is well to remember that just as new teaching software is constantly being developed, so is software to help researchers. Similarly, more and more researchers are sharing their experience in personal websites and blogs (see Part IV) as well as through more traditional publications (see section 15.6). Finally, not all research needs to record all data, and a narrower focus is sometimes desirable or even unavoidable. As Lemke said of his investigations into immersion games on the Web: 'We would like to understand class, gender/sexuality, cultural and subcultural differences in which games people play, how, and why; the kinds of meanings they make and feelings they experience; and what persistent learning effects result. But we need to take such an ambitious agenda one step at a time' (2006: 11).

Early decisions about the scope of your research (preferably a modest and well-defined one) is your best ally in succeeding with a CMCL project. In the following sections discussions of these issues are followed by summaries (in the form of checklists) which you might like to refer to when planning projects.

15.1   You and your participants' technical competence

Participants' technical competence may affect the content of the research or the process, or both. If the research is about mediation and how technological skills interact with teaching or learning, as in Lewis's (2006) reflective account of his experience as a novice online tutor, descriptive research designs will be needed that can accommodate the heterogeneity of participants' skills. On the other hand, if the chief concern is effectiveness in promoting acquisition, then homogeneous groups are needed (Felix, 2005). Homogeneity can be achieved by selecting participants on the basis of their technological competence or by ensuring that sufficient time is built in to train them. In section 1.6 we saw that Hubbard (2005) identifies the 'technological novice' status of learners or teachers as a factor liable to distort results. Such novice status receives little problematisation in the studies he has examined. However, as long as the tools are still not widely available to educationalists – tools such as voice-over-Internet systems with multimodal facilities – it should still be expected that many participants in CMCL projects will be relative novices, and you should allow for this as part of the ecology of the project.

Technological competence is required from researchers, as apparently trivial details can derail the research process. For example, one of the current authors was using screen capture video software to record a CMCL session but forgot to disable her screen saver. When she returned to her computer after a few minutes' absence, the screen saver had taken over and its images were being recorded, ruining precious visual evidence of the interaction in the virtual classroom.

**Project planning checklist 15.1: Participants'
technical training**

Option 1: training not provided

Has the research objective been examined for its compatibility with the use of a mixed ability cohort?
Has funding been allocated to pay for technical support for less confident participants, or will others be expected to help out (e.g. tutors, competent peers)?

Option 2: training provided

Has time been built in for training pre-experiment?
Will training be purely technical or will it include some application of
technical skills to content (i.e. learning to use the environment through
language activities)?
Has funding been allocated to write up the training sessions, their delivery,
the administrative support, the technical support?

15.2   An ethical framework for your project

Ethics is an important consideration, as ethical research is a legal
requirement and specified in an increasing number of countries. Even in
cases where a country has weak ethical protocols, international journal
publishers and book editors are unlikely to accept work that has not
received proper, documented ethical permissions. As we saw in section 4.2, methodologies requiring the detailed discussion of learner data
are widely used in CMCL research. This means that our field is one
where you as a researcher are potentially in a situation of power over
those whose behaviour you are researching. Considerations involve:

- Obtaining your participants' (or in the case of children, their
  parents') informed consent in writing, which implies your deciding
  exactly how much of the aims, methods and outcomes of the
  research they need to be made aware of before they sign the consent
  forms.
- Planning how to edit and present your data should some participants
  withhold permission while others give it.
- Adhering to the requirement not to coerce participants, even implicitly. For example, when a teacher asks students for permission to use
  the interaction for research purposes, the students must be reassured
  that withholding permission will have no negative bearing on how
  they are treated on the course.

15.2.1   Protecting participants from
misuse of published findings

Data protection has time-consuming consequences for the preparation of
studies prior to dissemination. If learner interactions were to be viewed as
purely linguistic phenomena, anonymisation would be a straightforward
matter of ensuring that participants' real names do not appear in the
transcripts. Even in such a simple case, however, questions can be asked
about what should be anonymised (surnames, given names, nicknames,

geographical or photographic data contained, for example, within the attachments that learners may have posted to the group as part of the task). Additional issues arise about the loss to researchers of sociocultural information consequent on anonymisation. In Figure 15.1, the screen image has had to be carefully edited for potential clues to identities, from machine-generated information (grey bars) to names and other personal details revealed by the participants about themselves or about others (black bars). Some of the information left unmasked could furnish clues to those who are in real life acquainted with the participants, yet blacking it out might injure the overall meaningfulness of the data presented. The researcher, working with the editor, will need to exercise common sense in these matters when it comes to publishing the work.

Finally, surface anonymisation might not be sufficient, as the meta-data (e.g. technical information embedded in a word-processed document) associated with documents may still bear traces of the identity of the participants. A further step may have to be built in to ensure that meta-data are anonymised as well, but advice from skilled colleagues should not be difficult to obtain should you require it.

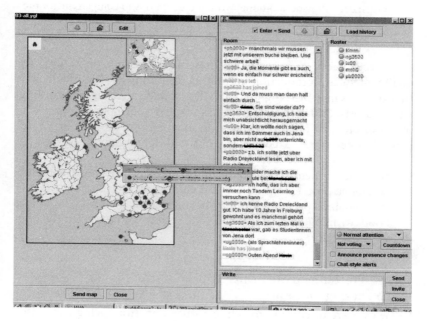

Figure 15.1   Anonymising screen images

Another challenge results from the fact that all CMCL environments, whether minimally multimodal (text chats) or more densely multimodal (videoconferencing systems and virtual worlds), provide researchers with observables that are not only linguistic but visual and auditory. Here are some of the issues involved in publishing material while protecting the anomymity of participants whose names, faces or voices are part of what the electronic environment provides by way of observables: sometimes the role of the visual data is ancillary, but in other cases it is central to the research question (e.g. where pixellisation of faces masks evidence of the way that participants communicate through proxemic means and body language). This duality is reflected in the next project planning checklist:

Project planning checklist 15.2: Anonymising images for publication

Is the name of the person, or the expression on their face an integral part of what is being discussed within the research?

No   Pixellise the name/face; mask the name.

Yes   Obtain participant's written informed consent, i.e. ensure that participants understand precisely what will be seen in the published image.
If consent is withheld, convert the data into another format, then edit out visual data. Provide additional written descriptions to compensate for the loss of information.

A special problem arises when presenting research on sound. The quality of voice reproduction in Internet telephony has become so good that speakers may be recognised on video soundtracks, audio recordings and podcasts. So far the problem has not been surmounted, and your only option is to obtain written informed consent from your participants. They need to be made aware of two different public settings:

1   conferences, where the data have limited exposure to a known audience;
2   conference proceedings online and conference websites, where your sound-enabled presentation may be made available for an indefinite period, to an audience wider than the cohort who attended the original conference.

15.2.2   Protecting participants' welfare during the project

So far we have discussed protecting participants from being recognised when your project results are disseminated. But as a researcher you also have a duty of care during the research project. We are talking here about protecting participants from possible harassment, which brings with it a need to ensure that environments to which learners are invited should be password-protected. This is because in public access sites there may be no or inadequate control over members' behaviour (see Jazwinski, 2001, on gender online; and the discussion of 'predators' in Cziko and Park, 2003: 25). Restriction to password-protected environments often means being limited to your institutional VLE, which may not be the environment that best supports your research focus. Again, a practical solution may be to try and negotiate password protection with public access site-owners for the duration of the project. There may be a charge for this service.

Project planning checklist 15.3: Ethical considerations

Your institution has a body overseeing the ethics of research:

Have you read your institution's ethical research policy?
Have you built in the time required to apply to its ethical research body and receive its response?

Your institution does not have an ethical research body. You need to:

Check your country's data protection legislation.
Decide what type and extent of dissemination you want to give to your findings.
Identify the different items that will need to be made anonymous.
Obtain your participants' written informed consent.

In both cases, if you are hoping to publish your work, you need to consult the journal or publisher's website to check on their own ethical requirements.

15.3 What practical consequences can you expect to face when researching home-based distributed learning?

Although some research takes place in school or university computer labs, distributed networking conditions are at the core of many projects and are challenging for researchers, as learners are connecting via the Internet from home, office, library or wherever they happen to be, possibly in different time zones. Issues may include the quality of your participants' equipment, the different configurations they may be using, or unequal learning conditions arising from the fact of their location as physically separate from the rest of the group and the teacher, perhaps in the family environment.

In synchronous environments, the quality of participants' equipment may affect:

- An individual: connecting from a slow modem may limit a participant's opportunities for communication and alter her/his performance.
- The group: if other participants are working from broadband access, there may be an issue of group dynamics.
  - some members may dominate the conversation because their audio or video connection is more stable;
  - the opposite might happen: those experiencing audio or video difficulties may monopolise the others' attention through appeals for support via other communication channels such as a text-chat tool running in parallel to the main speaking activity;
  - another reason for poorly-balanced group dynamics might be technical issues: busy networks in some countries at certain times may result in slower download rates and more frequent disconnections for those using modems.

The configuration of personal equipment may also introduce discrepancies. For example, variations in screen definition or browser setting may result in shared images looking different (paler/darker, blurred, cropped differently) on different participants' screens, with consequences for the verbal interaction. This may be detrimental to the talk, or serendipitously beneficial (see Lamy, 2004), but in either case contextual information extraneous to the interaction data will be needed if the researcher is to represent the interaction fully.

Other variations may be due to the fact that the group and the teacher are not co-located. Examples involve personal work habits and multi-tasking. Unknown to the researchers, participants may have several systems open on their desktop and be sending private messages or emails which the researchers will not be able to add to their data. This may be a distraction or significant as part of completing the task – for example, if a translation website is being accessed by a member working to help the group overcome lexical obstacles. Unless the learner explicitly refers to this activity, the researcher will not be aware of it and this will have research consequences (see Jones' (2004: 26) call for a broadening of the notion of context in CMC research).

The home environment and offline social factors are another possible cause of variation. Synchronous participants may be surrounded by activity, noise or family demands, impacting on the attention they are able to devote to the task and on the comparability of research findings with peers unencumbered by such distractions. In effectiveness research projects, such distractions have to be taken into consideration when accounting for the quality of performance.

In projects where it is important to minimise the role of such variables, standardising the subjects' equipment by loaning it to them, and their environment by locating them in a computer room, may be an option, but the former requires funding and the latter affects the ecology of the situation. In ethnographic projects, on the other hand, rich descriptions of communicative activity are required, including the fact that a subject may be simultaneously communicating with remote and with co-located interlocutors.

Project planning checklist 15.4: Networked participants connecting from home

---

Option 1: You decide to ensure homogeneity of equipment across the group through:

Loans: this assumes that you have already obtained funding for this purpose. Selection of participants based on the equipment they access to. In this case, have you:

- built in the time required to write and disseminate a minimum specification and to process responses from potential participants?
- devised a process for the selection, checking that they have correctly assessed their equipment, communicating with inadequately equipped but perhaps insistent volunteers whom you wish to eliminate?

Option 2: No selection, you need to:

- publicise your minimum technical specification well ahead of the project start date, as some participants may be willing and able to change their equipment, arrange to connect from a better-equipped friend's home or from a university laboratory.
- examine your research objective to identify issues that might be affected by the heterogeneity of participant equipment, in order to devise solutions if possible.

15.4   Guarding against the effects of automatic indicators

Computers display on screen and retain in their meta-data information that has been automatically generated. You may not have any influence over whether and where such information appears. This will affect your research at the ethical, content and management levels.

Ethically, the concern is the potentially ambiguous use that participants may make of telepresence indicators, i.e. icons or symbols signalling attentional statuses such as 'offline', 'online but away from the computer' and 'at the computer but not available to talk'. For example, participants may pretend to be away from the computer by clicking the 'away' icon, yet be listening in in order to deceive, tease or for other reasons. Or the 'deception' or disguise may be part of the task itself, as happens in avatar-based virtual worlds or in some MOOs. Whatever motivates it, anonymity on screen is not reflected in the systems-generated research data, which identifies log-ins. For researchers, the availability of information on the real identities of disguised participants has a procedural consequence: they must build in a step in the handling of the data in which pseudonyms and blurred images are reconciled with the original data and a log kept of the link between anonymous and non-anonymous files.

Automatic 'tracer' information itself may affect the content of the interactions in the project. One example is the impact on communication of 'message history' facilities, informing users which participant has opened, read, ignored or replied to a message. These facilities provide participants with information that may lead them to a way of constructing their own participation. For example, someone may refrain from opening a message from a co-participant with whom they do not wish to interact in order to avoid leaving a visible trace of their interest in the co-participant's posting.

Finally, as the organiser of the research you are not immune to the (possibly unwelcome) effects of exposure in those environments where your presence is revealed as an automatic effect of the system. Learners will be aware that they are being observed. The observer's paradox is one of the problems you have to address. The Open University explains the phenomenon in a textbook for researchers in sociolinguistics.

Concept 15.1   The observer's paradox

Do I tell friends and family, for instance, that I am looking out for differences between the speech of men and women? Might this result in them behaving differently? This common problem of research is known as observer's paradox and refers to the fact that people change their behaviour when they know they are being observed. If they know exactly what a researcher is looking for they are even more likely to subconsciously change their behaviour. Some researchers attempt to minimise intrusiveness in the hope that people will forget they are being observed or recorded. How far this is possible depends on the setting for the research and the method of collecting the data. It is, for example, hard to ignore a video recorder.

(Open University, 2005: 80)

Another difficulty arises from what Sarangi and Candlin (2003) term the participants' paradox. This refers to the activity of the participants observing the observer. If the system automatically displays information about what the observer is doing (e.g. by displaying the observer's name on parts of the screen where s/he is active), participants' attention may be drawn to the observer's activities, potentially changing their behaviour and thus skewing the research. Alternatively, the system can be configured to conceal the observer's presence, in which case the question becomes an ethical one. Both scenarios are troublesome, but again explaining the position to your subjects will mitigate – though not remove – the problem.

15.5 Summary

In this chapter we have looked at the issues that arise from the fact that your project involves human participants online. Unlike offline educational research, CMCL practitioner research requires you to concern yourself with the technical competence of all those involved. We saw that, like offline research, your CMCL research should be ethical, but the computer medium makes its own demands on your vigilance. Unlike most educational research, your subjects are likely to be dispersed, and we touched on the issues related to this situation. Finally, we turned to some ways in which the computer's tracer systems may impact on your research.

Further reading

From CMCL

Hubbard (2004). Noting that the typical language student has received little if any training towards using CALL and CMC tools for learning, Hubbard provides guidance organised into five clear principles. Although confined to the classroom situation, this chapter offers very practical considerations on learner training for a wide range of CMCL tools.

From other fields

CORINTE (in French). This is the site of a project by researchers into corpora of spoken interactions. Its pages on 'questions juridiques' offer much food for thought on participant rights and data anonymisation. http://icar.univ-lyon2.fr/projets/corinte/

Roger (2007). An open-access document by the Department of Linguistics at Macquarie University about ethics and power in educational research. http://www.ling.mq.edu.au/research/Navigating%20Research%20Ethics.pdf

Research Ethics. The open-access Open University page on research ethics, providing links to guideline documentation for human participant research. http://www.open.ac.uk/research-ethics/index.shtml

Hewson (2003). A brief article on Internet-mediated research in psychology. It contains useful generic advice about participant protection and researcher control in Internet-based surveys. A bibliography and a set of open-access websites for Internet-based questionnaire construction and administration are also offered.

16
Data in Practitioner Research

It is perhaps an intuitive thought that the more heavily technologised a research situation is the more of a team effort it requires, whereas projects dealing with simpler technologies (e.g. email) are more easily within the capacity of a well-organised individual. To an extent this is true, but you can still carry out small-scale research in the more multimodal media if your aims are well defined (see the introduction to chapter 15) and your expectations are realistic. Expectations can only be kept at a realistic level if you are aware of the demands of the overall environment, even if you do not have to work with all aspects of it. This is why in this chapter we review data-related issues that might affect teams or individuals.

16.1   What counts as data?

Researchers operating in non-electronic environments have to devote much time to producing videos and transcripts. Automatic recording, tracing and archiving make this aspect of CMCL researchers' work much easier than their non-networked colleagues', particularly those working with text-based data. Yet the technology cannot currently automatise all the processes. Furthermore, it brings with it additional processual issues. As an example of the first point, consider how audio and video data, once digitally recorded, have to be transcribed, to clear and exacting standards, as do recordings from face-to-face events. The second point relates to the multimodal nature of CMCL data. To capture the communicative actions of participants in modes other than the linguistic, researchers must be able to trace and synchronise new types of data. As McCambridge remarks in her study of language learning by the deaf in multimodal environments, in such projects a conversation may 'consist of a picture or webcam picture, text, smileys, animations, sent files, and links' (2006: np).

For example, researchers may want to keep track of which hot buttons and icons users clicked and in which order, or how users built up a shared text or drawing. In a virtual world using avatars, the researchers may want precise logs of how the users move their avatars around the space, and what elements of mimicry (nodding, smiling, blinking, etc.) they made the avatar perform. This not only necessitates the use of screen capture software, but also requires that the screen of each user be captured, as not all users see exactly the same sequence of actions on their screen. The users' posture and body language in front of the computer may also be of interest (see Garcia and Jacobs, 1999, for asynchronous CMC). In this case, again, video is required. If you are planning to use this medium, we recommend Mondada's (2006) study, which offers much to those filming classroom interactions as well as to the CMCL community.

As alluded to in the previous paragraph, research teams have different data processing (and collecting) priorities depending on their research orientation. This specialisation allows for the otherwise daunting prospects of processing multimodal events to be considerably mitigated, as long as sufficient time has been devoted to agreeing exactly what each researcher wants to focus on and to deciding precisely which traces, interactions and data should be collected (both during and after the project) for each team. Digital tools may allow for easy saving of online events in their raw form, but it is unrealistic and unnecessary to expect to process all the material.

To give an idea of the chain of data treatment that we anticipate a CMCL project would need to include, below is a list from Mondada (2005). She distinguishes between primary and secondary data. Primary data include recordings and documents ancillary to recordings; secondary data comprise everything to do with transcripts and meta-data. The crucial concern is to ensure that synchronisation between primary and secondary information is never lost, no matter how many transformations the data undergo during the research process.

To this we would expect to add a set of documents commenting on, and structuring access to, all of the files above, organising them into a coherent and legible whole which should be able to reconstitute for researchers, in as many ways as desired, information about the original experience.

Primary data

- Audios
- Videos
  in different versions according to their modes of:
  − digitisation
  − compression
  − editing (of various sources)
  − anonymisation
  − objects produced, consulted, transformed, used by participants during sessions (e.g. lesson plans, task instructions, preparatory papers, drawings, diagrams, summaries).

Secondary data

- Transcripts
  in different versions according to their:
  − granularity (rough vs. advanced transcripts)
  − anonymisation
  − alignment
  − annotation
- Transcript conventions
- Fieldnotes
- Meta-data comprising:
  − participant's descriptions
  − contextual and ethnographic relevant information
  − description of the recording set-up and processes (who was recorded, for how long, with which software and how edited)
- Authorisation forms

(Mondada, 2005)

Figure 16.1 Types of data

16.2 How should your data be presented?

As with data recorded in a classroom, a method has to be agreed for selecting the material to be collected. The recorded material can be listened to or viewed several times to draw out common priorities, or if you are working in a team, each researcher may view the data independently and negotiate priorities with their colleagues. But the main concern with CMCL data is that they are copious, and the more multimodal the environment, the more complex the issues concerning presentational choices.

In multimodal synchronous settings, it is important to ensure that the form and content are presented in accordance with the research aims, and ascribed to a time-frame that fits the real time-frame of the events. Matrices with columns are often chosen to represent simultaneous phenomena, sometimes with added incrustations showing still-shots of the action. Yet, as Thibault asserts, the layout may create a false impression.

Quote 16.1   Tables as a semiotic resource for representing multimodal data

The left-to-right visual orientation of the table is not without its consequences for the ways in which the makers and users of transcriptions perceive the relationships among the various components of the transcription and, by implication, of the text transcribed. [In the Western tradition of visual literacy], left is perceived as signifying **both** temporal and logical priority. That is, that which is placed on the left of the transcription is – probably unconsciously – doubly privileged on account of these organisational principles in the grammar of visual semiosis in Western cultures. Typically, transcribers place the verbal or linguistic component of the transcription on the left. If other semiotic modalities are referred to at all, they tend to be placed to the right of the verbal component.

(Thibault, 2000: 318; original emphasis)

Just as section 16.1 illustrated the importance of research priorities in driving the processing of data, here we show that they also drive the choices that you make about the presentation of your data for dissemination. For example, two different research projects might be using the same conversational data with different orientations, the first sociological, the second semiotic. In the first project, the priority could be 'emphasising the community aspect (if any) of a computer conference [, which] might mean focusing attention on the subjects playing an active role as senders' (Pacagnella, 1997: np). This orientation might lead to representations (e.g. charts) prioritising certain data automatically generated by message history tracing systems. In the second project, researchers could be interested in the communicative affordances of multimodal environments, with a focus on how a conversation is constructed through the parallel deployment of speech and text-chat. For them, a more useful representation would be a matrix showing which mode is used and when. Each of these orientations would determine choices at every stage of the chain of data processing: transcripts, annotations, labels and tags, analyses and associated tools.

Project planning checklist 16.1: Presenting the data for dissemination

1 Use the checklist to consider dissemination within the project team during the project.
2 Then use it to consider dissemination to the rest of the research community post-project.

	Some pros and cons	Pros and cons for your project
Textual transcripts	*Pro*: availability of pre-existing transcription protocols; compatibility with existing automatic tools for qualitative analysis. *Con*: limited ability to represent multimodal phenomena.	
Matrices	*Pro*: familiar to and thus legible by the research community. *Cons*: not a neutral way of representing data (see Thibault, 2000); representation of synchronicity of actions not always accurate; can be very detailed, hence not easy to use for digital presentations or print illustrations.	
Screen-shots or stills from screen videos	*Pro*: attractive and eloquent as to what the experience of participating in the project was like (sparing the researcher the need to write lengthy descriptions). *Cons*: cannot isolate actions by type, because bound to chronology; file format may cause problems; very colourful, hence sometimes not acceptable to print publishers.	
System logs	*Pro*: objective data; easy to convert into charts. *Con*: may require input from systems staff in order to retrieve and interpret them.	
A mix of all four	*Pro*: a flexible way of representing data. *Con*: complexity.	

16.3   What is a corpus and do you need one?

Due to factors beyond the researchers' control, data may become inaccessible; for example, the institution that hosted the event may cease its commitment to the original platform and alternative storage may not support ways of keeping data linked to their original context. Data may be buried, in a proprietary format, within the technological environment. Each screen capture video produced as part of the research process only provides evidence of what happens on one participant's screen, yet the entire group interaction needs to be restored to its original multidimensionality for a full analysis.

The answer is to turn collections of data into contextualised corpora. Such corpora include not only the data that are the direct output of learner activity on online courses, but also their context, i.e. information about the pedagogical and research settings. Creating such a contextualised corpus involves a reflection on the work of transcribing, annotating and analysing multimodally. In other words, to qualify as a corpus, a data collection comprised of primary and secondary data (see Figure 16.1) has to be accurately synchronised and exhaustively documented. Some of that documentation inheres with data integral to the recordings and transcriptions (e.g. a log of the start-time of a conversation) and some is provided by server managers (e.g. records of connections and disconnections).

16.4   How can you store and preserve your data?

Having decided which part of your data needs to be processed, the next question facing you is storage. Synchronous CMCL data are voluminous, particularly if they include audio recordings. For an illustration, we looked at a 90-minute session from one of our recent projects, in which participants talked and text-chatted simultaneously while sharing graphic tools. The size of the files generated by these graphic tools only totalled about 5 MB, and the text-chat files were about 3 MB, but the audio output archived as a .wav file was 447 MB. The session was also video-recorded with screen capture software and archived as a file in .avi format (i.e. soundless video) of 200 MB. Finally, a file bringing together video and sound was assembled in order to reconstitute for researchers an experience as close as possible to the real event, resulting in an additional 300 MB. The funding and purchase of external storage space (disk drives and CDs) was a vital stage in the planning for that project.

You also need to plan for the shelf-life of your data. As mentioned above, project files are held on institutional servers which are liable to undergo changes which may not be congruent with the interests of the researchers. You will need to find out about such plans in advance and take steps to protect your data. This may involve transferring them to other servers or to external devices.

Project planning checklist 16.2: Storing and preserving data

Have you checked the likely life expectancy of the platform at your institution?
Have you checked (e.g. by searching the Web for news items on software companies) the likely availability of the software you are planning to use over the years that research will continue on the project?
Have you obtained funding to buy independent storage devices and bought them?
Have you researched the likely impact of transfer from original platform to independent device in terms of data structuring?

If there is an ICT policy group or department at your institution, that should be your first port of call. If there isn't, or if the answer doesn't fit your plans, you may need to consider using Open Source conferencing or tutorial packages. Then you will be free of the risks that might follow decisions made by your institution, although you will probably be denied technical support from them. If choosing an Open Source platform, you will still have to be aware of possible threats to its viability. To ensure that you are updated, you could put questions to online networks and/or contact the platform administrators directly via their website (for Open Source packages, and professional networks, see Part IV).

16.5   What automatic tools are available for analysing CMCL data?

With the abundance of data generated by CMCL, automation of the processes is desirable for both quantitative and qualitative analyses. Part IV lists Internet addresses linking to several automatic analytical tools and in some cases to user assessments of their usefulness. Here we touch on some we have found particularly useful.

If you need a statistical package, you may find that SPSS™ is well suited to the needs of educational research, but you may also wish to inspect other statistical packages such as Instat or Statistics for The Terrified, which offers a brief and a more in-depth demo (both free to download). As to software for the analysis of qualitative data, it has been available since the early 1980s (and is known by the acronym CAQDAS – Computer Assisted Qualitative Data AnalysiS). The present researchers have used N'Vivo and found it useful, but as change is rapid in the research software industry, you would be well advised to visit (and regularly revisit) one of the websites describing the types, uses and shortcomings of CAQDAS tools. One such website has been compiled by Gibbs, Fielding, Lewins and Taylor – http://onlineqda.hud.ac.uk/index.php – where you will find hotlinks to proprietary websites.

According to Gibbs et al.'s web resource, there are four types of tool:

1 'Text retrievers' search for items, but cannot offer you any help in coding them.
2 'Code-and-retrieve'. The more recent packages allow you to add memos, diagrams, charts, concept maps and ways of refining the coding system itself. More recently still, software has been produced that can carry out these functions based on audio and video inputs.
3 Concordancers. These generate lists of words (keywords) taken from a body of texts (corpus), displayed in the centre of the page with the contexts in which they occur (based on some measure of how large the context is, e.g. 29 characters to the left of the keyword and to the right of the keyword). Many types of text-based analyses are supported by concordancers, in language teaching (Tribble and Jones, 1997; Lamy and Klarskov, 2000) and in generic social science and education research.
4 Data converters. Two types exist: software that converts written data from one format to another, and optical and voice recognition software that converts audio (and even video) into text.

The majority of these tools are built to handle linguistic modes and are familiar to social sciences and educational researchers, so your institution may already have a licence for their use. Checking these arrangements and whether they apply to you is a priority if you are embarking on an analysis of written CMCL.

Software constructed to cope with multimodal data, such as the concordancer MCA3 (Baldry et al., 2006), which allows the coding and retrieval of film sequences, was developed as a resource for face-to-face language teachers and learners. Its application to both online language teaching and CMCL research is still a potential rather than a reality. However, one recent development worth watching is TaSync, specifically designed to assist coding, retrieving and organising of multimodal

CMCL data (Betbeder, Reffay and Chanier, 2006). Adaptation of generic CAQDAS software to the needs of CMCL research is, we suggest, a research and development agenda worth pursuing for the future. See Part IV for further suggestions about Internet-based information on automated data analysis tools, including text analysis tools for dealing with languages other than English.

Project planning checklist 16.3: Should you use software analysis tools?

Different packages have different qualities. What can software can do? What do you mainly need the software to do for you?

> Structure your work.
> Enable access to all parts of your project immediately.
> Help to explore your data through word or phrase searches.
> Create codes and retrieve the coded sections of your text.
> Search for relationships between codes.
> Provide you with writing aids: memos, comments and annotations.
> Allow you to print hard copy or export to another package.

This is what the software cannot do. Have you planned how and when you will carry out these functions?

> Analytical thinking (though it can support the thinking).
> Coding (some software supports automatic coding, but you will need to check the results).
> Reduce bias or improve reliability (though it can support such improvements).
> Calculate statistics (though some software offers minimal quantitative support functions).

(Based on Lewins and Silver, 2007)

16.6   Summary

In this chapter we have discussed how to identify the data you need. To aid your planning, we have proposed a classification of data types and showed that the problems linked to choice of data are echoed by the choice of representation of data. We have discussed collecting corpora of data and the issues related to digitally storing it. Finally, we have offered a brief introduction to automatic tools that can help you with different stages in your project.

Further reading

Séror (2005). A short paper offering a condensed history and overview of CAQDAS tools, and a discussion of their advantages and disadvantages.

17
Some Possible Practitioner Research Projects

So far in Part III we have discussed the nature and structure of practitioner research projects as well as the practical conditions under which such projects should take place (relating to participants and to data). In this chapter, we bring this information together to create six templates for small-scale practitioner research projects about CMCL. We have designed these so that you can run the projects with as limited an outlay of resources as possible, either on your own or with a very small team. The six projects respond to several of the issues discussed in the book. Thus each one:

- tackles a research context signalled in Part II as needing further probing;
- uses methods and tools identified in chapter 14;
- requires some of the ethical precautions advocated in chapter 15;
- involves some of the technical procedures outlined in chapter 16.

17.1 How to use the project templates in this chapter

In the rest of this chapter, you will find a template for each of the following:

Project 1: Researching videoconferencing
Project 2: Researching learner identity
Project 3: Researching collaboration
Project 4: Researching emerging technologies
Project 5: Researching teacher training
Project 6: Researching online tasks

For all six projects, we have chosen research questions arising from the alliance of pedagogy with technology. However, there are two ways in which you can expand on our offering and generate a multiplicity of other projects.

1 Recombine the elements of two templates. For example, you could use one template but modify it by inserting a task from a different template or by changing the participant profile (e.g. by using teachers instead of learners). Or you could adapt a case study to research a different topic (e.g. the case study in Project 2 could be used to research one of the emerging technologies in Project 4).

2 Start from tasks designed for offline language learning and teaching and adapt them for CMCL investigations. Bygate and Samuda (2007) offer a wealth of possible offline research projects centring on tasks. Adapting these projects for research into CMCL can be a fruitful strategy, providing the essential condition of CMCL research is observed, that is, you should always be able to answer in the affirmative the question: 'If this project was carried out online, could it generate research questions specific to the fact that the learning and teaching are technologically-mediated?'

17.2 Six project templates

Project 1: Researching videoconferencing

Why do this project?

The first question that we are often asked when we talk about our 'distance' work is: Do you teach by videoconference? There is an expectation that our methods should include the video aspect, yet, as we explain in section 12.4, many questions about video remain unanswered because they are not asked. However, projects designed to answer them can easily be set up now that video technologies are accessible to all those with a broadband connection. Although such tools are not specifically designed for language learning, language learning tasks can be delivered using them.

What is the main context of your research?

The affordances of a generic video-based environment for language learning.

What are your research questions?

The main question – What role does video play in language learning and teaching online? – can be broken down into two sub-questions:

1 Learning: what perceptions do learners have of the advantage of the video image when being taught via video teleconferencing?
2 Teaching: what specific advantages accrue to teachers from the video mode of video-based teleconferencing?

How should you determine your priorities in setting up the project?

1 The visual dimension of the project should be the first priority and be clearly reflected in the design of the task.
2 The second priority is to define learning outcomes, avoiding the pitfalls of complexity. For example, prioritise one outcome among several possibilities such as cultural learning, vocabulary development or oral fluency development.

Which procedures should you follow?

Participants:
- Save on resources by bringing together participants who will provide information on each of the two research sub-questions, i.e. a teacher group and a learner group. For example, you might ask (a) colleagues from your institution, (b) pre-sessional teachers from a training institution or (c) postgraduate or mature undergraduates (as Mrowa-Hopkins, 2001, did in respect of forum-based work) to teach learners. An even number of learners works best for this project.
- Number of participants: minimum three teachers and six learners; maximum: depends entirely on your own energy and resources.
- Reasons for the learner group to agree to participate: added practice, opportunity for individual attention from a teacher. Alternatively, make participation in this project an assessed part of the course.
- Reasons for the teacher group to agree to participate: opportunity to watch their own performance privately, or if preferred with a 'critical but friendly' colleague; opportunity to receive for their own use a collection of tasks created by other pairs.

Equipment:
Choose a free tool such as MSN messenger® which incorporates audio and video, and has a facility for audio and visual recording of sessions. Check that this tool is allowed for institutional use and that it will not

be blocked by firewalls. Ensure a supply of webcams (minimum: one per teacher and one for two learners working together at a computer).

Set-up:
- Number of 45-minute sessions: minimum six; maximum: depends entirely on your energy and resources.
- Each tutor works with two learners. The learners will need to be in your institution's computer lab, so you can supervise the recording of the sessions. The teachers can work from their institution or from home.
- A minimum of one face-to-face debriefing session should be organised for both the learner and the teacher groups.
  - Aims of the learner group debrief: to collect evidence towards research question 1 and to increase learner reflectivity.
  - Aims of the teacher group debrief: to collect evidence towards research question 2, increase teacher reflectivity and pool the activities and their feedback for post-project use by teachers.

Task:
Ask each pair in the teacher group to devise a task to last for a specific duration online (e.g. 45 minutes). You could specify general parameters for the task to ensure some degree of comparability, and you should brief your teacher pairs to make use of the visual aspect of the setting in their tasks. The pairs should design the tasks independently of each other. This will create variety and maximise the benefits to individuals at the point of debriefing. Each teacher will deliver to two learners the task that s/he co-designed.

Data types and collection:
- Session recording: see section 16.4 for precautions to take in the collection and storing of large audio and video files.
- Questionnaires: both MCQ and open-ended.
- Interviews with members of each group.
- Researcher notes from the debriefing sessions.

Data analysis:
The learning research question is about perceptions. A qualitative analysis of interviews, of open-ended questionnaires and of debriefing session outputs by learners will provide language to be analysed using DA methods (see section 4.2.2), possibly with the help of a simple, free piece of semantic analysis software such as Tropes (see section 16.5 on software for qualitative analysis, and section 18.9 for Tropes).

The teaching research question can be approached through the quantitative and qualitative classification of teachers' comments in interviews, questionnaires and debriefing notes. The analysis will be complemented and empirical examples or counter-examples identified through a review of the session recordings.

Project 2: Researching learner identity in virtual worlds – a case study

Why do this project?

CMCL not only has an impact on how students work with others (see Project 3) and on how, as a consequence, the social identity of a group develops, it also raises questions of individual identity. In virtual worlds, avatars give users a virtual graphical presence in a 'body' that may be a relatively realistic human image or a fantasy being.

What is the main context of your research?

As we have indicated, the introduction of a new cultural tool has a transformative effect on what users do. The main context of this project is to try to understand the effects of using avatars and how this functionality can be exploited beneficially in language teaching.

What are your research questions?

- How do learners use avatars in virtual worlds?
- Do they use them to develop an identity? If so, how?
- What are the affordances of an avatar?
- What does this embodiment mean for interaction with others? Does it, for example, encourage role-playing?

Which procedures should you follow?

Set-up:
How you proceed will depend on whether you are planning to use a closed environment that restricts access to a group of users (for which you usually have to pay a licence fee) or a free and public platform such as Second Life, which has recently attracted the attention of some educationalists (e.g. Stevens, 2006).

Task:
Draw up a task which will work in the virtual environment you have chosen. If your students are working together in a closed group, you could consider a discussion in the form of a role-play that links in

with the syllabus your students are following. If you are using a public environment, you could ask them to investigate a cultural topic.

Participants:
A case study focusing on one student and his or her development during the project would allow you to limit the data to a workable size while still giving you usable findings.

Methodology:
As is apparent from sections 4.3 and 6.4, a topic such as identity is best studied using qualitative ethnographic methods, which are well suited to accounting for the rich picture that identity presents.

Data collection:
As the development of identity is a process that happens over time, record an initial session as well as a session towards the end of the project, using screen-capture software. You may also want to collect material created in the virtual environment. A critical event recall session, done in retrospect, using one or two of the recordings may help you get closer to the student's perspective. Choose a suitable session and encourage the student to reflect on their actions and record their comments by, for example, video-recording the session. A final interview will help you draw together the different strands of the project, that is, the original online sessions and the student's comments during the think-aloud session.

Data analysis:
What avatar has the student chosen? Examine whether and to what extent the character of the avatar influences the student's interactions with others in the online environment. The think-aloud protocol (see section 14.3) allows you to find out more about why the student behaved in a certain way and what the functionality of the avatar actually afforded the student in this situation. Discourse/conversation analysis may help you analyse these data (see section 4.2). You might also triangulate the data gathered during your observations with the data gathered in the interview as well as the think-aloud protocol.

Resources:
If students are not familiar with the technology, time needs to be allocated for training.

In terms of technical resources you will need a screen-capture tool (see chapter 16) to record the online sessions and enough server space for storing the files with the recordings. In order to carry out the critical event recall you will need to video-record the student sitting in front of a computer, commenting on the recorded online sessions.

Project 3: Researching collaborative learning

Why do this project?

Although the idea of collaborative learning as a pedagogically desirable outcome of constructivist pedagogies in online settings is widespread in the literature of CMCL, it is clear from Chaptal (2003), O'Dowd (forthcoming) and from section 8.4 that this requires further probing. Henri and Lundgren-Cayrol (1997) and Mangenot (2003) identify task types appropriate to collaborative learning. Among these are:

- mining of L2 resources (for production of a collaboratively negotiated object);
- critical analysis of L2 resources (for production of a collaboratively-negotiated object);
- debating in L2 (for publication of a collaboratively-prepared document);
- problem-solving in L2.

We offer a project idea for researching collaborative learning, using the first of these tasks. The shape of the project is such that one of the other tasks, or two in combination, could be used for the same research purpose.

What is the main context of your research?

Research on suitability of particular pedagogies for online language learning.

What are your research questions?

The main question – Does collaborative preparation of an 'object' (e.g. a poster, questionnaire, podcast) ensure that collaborative language learning has taken place? – can be divided into two sub-questions:

1 What do my data reveal about the relationship between collaboration and learning?
2 What do my data reveal about the relationship between collaboration and *language* learning?

How should you determine your priorities in setting up the project?

All aspects of the research, whether based on self-report or elicited from observation or log data, need to focus on the process whereby participants achieve the final outcome.

Which procedures should you follow?

Participants:

Minimum of three learners who will work individually, and six learners who will work in three pairs or two trios. Minimum of one 'evaluator' per individual and one per pair or trio.

This project should be a volunteer experience for all participants, as its technological dimension needs to be a motivator rather than an inhibitor. Finding volunteer evaluators of the outcomes prepared by the learners will add to the ecological value of the research.

Equipment:

Access to the Internet for all participants.

Set-up:

Individual participants and pair or trio participants are given a time-frame (e.g. a fortnight or month) in which to do the task and produce the outcome. All questionnaires and interviews are carried out after the participants' tasks are completed.

Task:

The same task is set for one or more individuals and one or more pairs or trios. The design consists in learners exploring a document of interest to them (e.g. a Web 2.0 site of their choice – YouTube, Flickr, MySpace, etc.), then performing an action related to this site (e.g. uploading a file or creating a link to the site) and finally producing an outcome. This outcome is an L2 account of what they did and why, for the benefit of novice peers who wish to imitate these actions. Novice peers can be imaginary or (preferably) real. Individuals are asked not to confer with peers. Those working in groups are asked to collaborate within their pair/trio. All are told that they may produce the outcome as a Word document or as any other appropriate form of dissemination.

If volunteer novice evaluators have been recruited, they will be asked to use the prepared outcome as a guide that should be able to help them imitate the actions of the learners. Evaluators then report on their success or otherwise to the researcher.

Data types and collection:
- Learners' L2 account of what they did and why.
- Questionnaires or critical event recall interviews collected after the event, focusing on the process by which learners produced their outcome.
- Interviews of the volunteer evaluators.

Data analysis:

The first stage of analysis is a comparison of the individual learners' L2 outcomes and of the pairs'/trios' outcomes. These could be couched in terms of:

- L2 quality and validity of the outcomes as an account usable by a novice (language learning-oriented data, research question 2 above);
- the form of the outcomes: did groups use more multimedia resources than individuals to present their outcome? (generic learning-oriented data, research question 1 above).

You might also triangulate this analysis with questionnaire responses and interviews.

Project 4: Researching emerging technologies

Why do this project?

A reason for looking into emerging technologies might seem obvious because they are new. As chapter 13 has shown, little research has been done in this area. However, there is a well-known danger of seduction of practitioner by new gadgets (the *Wow!* factor), and this project is designed to guard against this by retaining the pedagogical perspective at all times.

The project is what might be called an informed try-out, and could be adapted for blogging, podcasting, using wikis or PDAs, and other emerging communication devices. In the simple try-out suggested, you will only be able to gain an initial feel for the answer. For a deeper understanding of the affordances of the tool, repeat the project with different tasks.

What is the main context of your research?

Research on the affordances of tools.

What are your research questions?

Are there particular learning benefits that only this tool can deliver, or that this tool can deliver more easily (cheaply) than other tools?

How should you determine your priorities in setting up the project?

Your priority in all aspects of the project will be repeatedly to ask yourself the following:

- Can the learning task be done with other technologies or has this tool a unique role to play in getting the task done?
- Could the tool be better used carrying out a different task?

Which procedures should you follow?

Participants:
You and one (minimum) or three (maximum) colleague(s). All should be language teachers with an L3 in common.

Equipment:
- For all: personal audio-recording device.
- Access to the Internet for all participants (if researching blogs, wikis or podcasts). One relevant mobile device per participant (if working on mobile technologies).

Set-up:
Participants will work with a partner, in pairs. To minimise complexity, only one of the technologies will be explored, but similar projects can be organised later around other technologies. If a single technology is explored, the project can be carried out over a short period (e.g. a week) and the conclusions shared more meaningfully during the final participant debriefing session.

Task:
The task is a reflective narration by each participant, which will later be contrasted with a narration from their partners. To ensure comparability with what learners normally experience when asked to use a new tool in L2, the participants (professional users of L2) agree to work in a language that they know less well, such as a shared L3.

The reflective narration is the story of your and your partner's use of a blog, wiki, podcast or mobile device for the completion of an L3 task. The account should express the participants' positive and negative views on using the technology to do the L3 task. The L3 task can be simple, but should have a real outcome that can be published to all the participants. As the participants are all teachers, there should be no shortage of ideas, but here are some examples, in increasing order of technological demand:

- Researching wikis: your job is to add an entry to an L3 wiki (for added interest, this could be a topic of professional relevance to you and your partners). Your partner's job is to dispute, complete or in some way edit your entry.
- Researching blogs: your job is to create a basic blog in L3. Your partner's job is to help you by finding examples of attractive authentic L3 blogs for you to imitate. To achieve this, you both communicate in L3 via a forum, chat, email, mobile phone or other non-face-to-face communication medium.

- Researching mobile devices: with your partner, choose a local visit to make. This has to be a real visit, but depending on the L3 it could be to a shop (e.g. selling Italian goods), dance club (e.g. specialising in Latin American dance), natural setting (e.g. Japanese garden), religious building (e.g. mosque, temple) or whatever is accessible to both you and your partner and has some connection with the culture of your L3. Your and your partner's job is to make the visit together, document it on your mobile devices in L3 and use this to prepare a multimedia presentation advertising the visit's L3 learning potential to your colleagues.

Data types and collection:
- The reflective narration can be collected by recording a think-aloud session onto an audio file, then writing up the salient information in a report. Each participant should do this for his or her own session. If you prefer to ensure some uniformity of response, you might like to create a template for your colleagues to fill in as they choose what information to extract from their think-aloud session. This only really becomes necessary once the number of participants exceeds, say, five. If you are working on mobile technology, you may not wish to walk around while 'thinking aloud', so a variant consisting of note-taking immediately after your session may be more appropriate.
- One face-to-face debriefing with all pairs of partners will be useful for complementary information on perceptions about tool use. Record this session.

Data analysis:
Even with a small number of participants, the think-aloud sessions and the output of the debriefing will create a large volume of feedback to the researcher. Try to reduce this to, say, a set of tables, for example of advantages and disadvantages for learning and teaching in general, or of the relationship of tool and task.

Project 5: Researching teacher training – an action research project

Why do this project?

Online teaching of languages still faces a predicament: despite the wealth of materials and the willingness of teachers and institutions to offer online courses, there is still a dearth of high quality training to teach online (Stickler and Hampel, 2006). Innovative approaches to the learning and teaching of languages, especially through the use of technology, necessitate thorough, ongoing training programmes.

What is the main context of your research?

Professional development, in particular effective teacher training.

What are your research questions?

- What professional development are teachers offered in my institution?
- Is this professional development effective?

Which procedures should you follow?

Set-up:
This project assumes that you (or your institution) already develop teachers to teach online.

Participants:
Participants include those responsible for the professional development programme as well as the teachers who undergo this programme.

Methodology:
Qualitative methods are most suitable for such a project. However, if you are involved in the development of more than ten teachers, quantitative methods will usefully complement the qualitative ones.

Data collection:
Collate information on the training programme available in your institution to develop teachers' online skills. This may be restricted to pre-course training or include ongoing staff development and opportunities for critical reflection. If you do not offer the development programme yourself, interview those who are responsible for it, find out what the objectives are and how they are implemented in the programme. Observe some of the programme sessions and collect written evidence, such as forum contributions or peer mentoring documents. Interview at least two or three teachers in order to find out if and why they found/did not find the programme useful. Do so at two points: once the initial steps of the programme are completed and once when they are actually teaching on an online course.

Data analysis:
Analyse the training programme: What level of training does it provide? Is it limited to the mastering of the technology, or does it help teachers acquire pedagogical skills? (see online teaching skills in chapter 5). Is the support that teachers are offered restricted to initial training or is it ongoing? Are teachers encouraged to share their experience with their peers and become reflective practitioners? How do they perceive the training programme? Do they find it useful? Can any of the elements be improved? (see sections 5.3 and 5.5).

Resources:
As this is an action research project (see sections 14.1.1, 14.1.2 and 14.5) it presupposes that any insights into how the training programme can be improved will be implemented in future years.

Project 6: Researching online
tasks – problem-based learning

Why do this project?

Although the potential of virtual environments for problem-based learning is beginning to be recognised in a number of educational areas (see chapter 5), it has not been examined widely in the context of language learning.

What is the main context of your research?

Sociocultural approaches to learning (see section 2.2), which support the active construction of knowledge through problem-based tasks.

What are your research questions?

- What kinds of tasks are suitable for online problem-based language learning?
- What are the conditions for problem-based learning to be successful?

Which procedures should you follow?

Set-up:
You need to consider the online tools your students are going to use. In the context of problem-based learning a wiki may be useful for bringing together material and writing collaboratively. Your learners will also need an environment that allows for synchronous communication. Your institution may have a virtual learning environment featuring a portfolio of synchronous tools (e.g. written or audio-/videoconferencing) and asynchronous tools (e.g. fora, wikis or blogs). If this is not the case, you can use tools that are freely available (see section 18.9). Ensure that you have access to the virtual environments, students' spoken conversations are recorded and any texts are saved.

Task:
You need to devise a task that poses an authentic problem for students to solve. This could be related to their life at school/university generally or to their language studies specifically (e.g. connected with their term/year abroad, if your educational system includes such an option). For task implementation, you may find it useful to adapt the process

that Abdullah (1998: np) suggests for problem-based language learning tasks that are carried out face-to-face:
Learners, in their own groups, collaboratively

- generate working ideas or possible solutions (e.g. write a petition, suggest alternative measures, form volunteer student patrols, survey students' views and present them);
- identify available information related to the problem (e.g. school policies, sample petition, sections of the school most vandalised);
- identify learning issues (things they need to find out, e.g. survey formats, how to form patrols, what other schools may be doing);
- identify resources to look up or consult (e.g. home-pages of other schools, friends in the police force, sample survey);
- assign tasks to the various group members (i.e. who is responsible for working on each learning issue);
- gather information (e.g. visit websites, interview students and community members, draft a petition);
- propose solution(s).

Decide which parts of the task students are going to carry out online. This may include all activities or only the later ones of gathering information and proposing a solution.

Participants:
One or several smallish groups: 5–8 students.

Methodology:
An ethnographic approach using qualitative methods (see sections 4.3 and 6.4) is most suitable.

Data collection:
You need to focus on the process and the product. In order to trace the process, record all synchronous sessions and log asynchronous communication that goes on in the execution of this task. Collate all documents created by the students. You may also want to find out what students thought of the task, perhaps by using a focus group. Also think about what the experience was for you as a teacher – for e.g, what kind of support did you give?

Data analysis:
Examine the outcome. Have the students completed the task and solved the problem? If they have, analyse the process and see how they went about it and how they used the online tools. If they have not, try to

identify what went wrong. Use the findings from the focus group to see what the students thought of the experience. How much support did they need from you?

Resources:
Depending on the number of students and the extent of the task you may get a wealth of data which will need to be stored and analysed. If you use interviews or a focus group, you will need to transcribe and analyse them.

Part IV
Resources

18
Resources

18.1 Introduction

This final part offers a list of mainly online resources. Elsewhere we have shown how CMC as a field in general can be of help to those working in CMCL, so the list includes both these domains. Similarly, some generic resources (such as the Web Style Guide) may be of practical use to those designing materials to be integrated into a CMCL project. If we have found them helpful, we have included these.

The nature of the Web is such that characterising types of sites for inclusion posed a problem as there are many overlaps. For example, a portal may be dedicated to tools, or a professional blog may include an online bibliography. So although we have divided the resources into broad categories according to the main function of each site, and organised the sites according to the categories shown on the resources map below, our categorisations are not hard and fast. Following hyperlinks, particularly in the biggest category (information centres and portals), will often lead to valuable resources in a different category. Where we have selected links in main sites, we have indented them.

Resources map
Blogs
Information centres or portals
Online bibliographies
Online books
Online journals
Online newsletters

Professional organisations
Tools and practical support: free
Tools and practical support: pay-to-use
MOOs and virtual worlds
Video-streamed talks and other free educational sites
Wikis

18.2   Blogs

Humlab Blog

Run by students at Umeå University's humanities lab, a 'a place where the humanities, cultural studies and modern information and media tecnology can meet and work together, both in real and virtual terms'.
http://blog.humlab.umu.se/

ICT4LT Blog

This replaces the bulletin of the ICT4LT online resource (see section 18.5)
http://ictforlanguageteachers.blogspot.com

Technorati

The search engine for blogs, which you can interrogate with an advanced query form by keyword (try 'academic' or 'language learning') or browse via their blog index.
http://www.technorati.com/blogs

Therese Örnberg-Berglund's academic blog

Therese Örnberg's friendly blog discusses her research on virtuality and conversational multitasking.
http://blog.humlab.umu.se/therese

Telecollaboration blog

Robert O'Dowd's telecollaboration blog focuses on intercultural exchange.
http://dfm.unileon.es/telecollaboration/

18.3   Information centres or portals

Action Research

A site from Graduate School of Education at George Mason University (Virginia, USA) entirely devoted to action research.
http://gse.gmu.edu/research/tr/TRaction.shtml

Center for Language Education and Research (CLEAR)

This portal, based at Michigan State University, promotes the teaching and learning of foreign languages in the United States. It covers face-to-face activities, as well as web-based ones.
http://clear.msu.edu/clear/

Computer Assisted Assessment Centre

This website is the outcome of a now completed TLTP project. Some of the links may not be maintained, but it still contains many valuable resources and links to bibliographies, articles and other related websites.
http://www.caacentre.ac.uk/

December Communications

December Communications Inc. is a web-based communications company with a primary focus of offering online publications related to Internet use, reference, development and understanding. Hundreds of links to CMC (rather than CMCL) information and the blogosphere.
http://www.december.com/works/central/

Deutscher Bildungsserver

Contains a list of e-learning portals in German.
http://www.bildungsserver.de/zeigen.html?seite=1561

EducNet

Dossier (in French) on online collaborative learning.
http://www.educnet.education.fr/dossier/eformation/modularite5.htm

Edulinks: telelernen und internet in der lehre

A non-commercial portal with articles, links, books, conferences and software related to online learning in tertiary contexts.
http://www.edulinks.de/

Embedding Learning Materials

A website devoted to embedding learning materials into ELT. This is a well laid out website for learning and teaching professionals, and for developers working to promote innovation and best practice in education, where you will find a wealth of practical materials to download, designed to help you. The site is designed for ELT but has wide applicability to other languages. Last updated in 2003 so some of the links are no longer live. All those listed below were live at the time of writing (2007).
www.elt.ac.uk/materials.htm

http://www.elt.ac.uk/ELT%20documents/materials/Bibliography.pdf
Links to a generic bibliography of practitioner guides, texts for teaching and learning, designing learning resources, research-focused and theoretical articles, and articles on institutional and policy implications of embedding learning technologies.

http://www.elt.ac.uk/ELT%20documents/materials/journals.htm
Links to a range of generic journals on learning and technology.

http://www.elt.ac.uk/ELT%20documents/materials/conferences.htm
Takes you to a list on generic conferences interested in papers on CMC.

http://www.elt.ac.uk/ELT%20documents/materials/web%20sites.htm
Links to a portal of websites full of ideas for using CMC (and other technologies) in higher education, some of which require log-in identification.

http://www.elt.ac.uk/ELT%20documents/materials/resource-eval.pdf
A simple template that can easily be adapted to CMCL. Useful for keeping private notes on technologies encountered, or as a checklist or starting point for a user questionnaire.

http://www.elt.ac.uk/ELT%20documents/materials/evalguide.pdf
This evaluation guide provided many of the ideas for chapter 14. The online version of the guide is much more detailed.

http://www.elt.ac.uk/ELT%20documents/materials/evalplan.pdf
and http://www.elt.ac.uk/ELT%20documents/materials/ARplan.pdf
This planning resource takes the form of two simple templates to help keep track of your planned actions, and at the same time check the quality of planned interventions.

http://www.elt.ac.uk/ELT%20documents/materials/obsform.pdf
A teaching observation form that ensures you have thought of the most important issues, and helps brief your observers.

Ethics of human research

CORINTE
This is the site (in French) of a project by researchers into corpora of spoken interactions. Its pages on *questions juridiques* offer much food for thought on participants' rights and data anonymisation.
http://icar.univ-lyon2.fr/projets/corinte/

Navigating Research Ethics
Roger (2007). An open-access document by the Department of Linguistics at Macquarie University about ethics and power in educational research.

http://www.ling.mq.edu.au/research/Navigating%20Research%20
Ethics.pdf

Research Ethics

The open-access Open University page on research ethics, providing
links to a number of guideline documentation for human participant
research.
http://www.open.ac.uk/research-ethics/index.shtml

Galanet

This European-funded project (in French, Italian, Castilian Spanish,
Catalan and Portuguese) aims to study mutual comprehension among
speakers of the Romance languages. The site reports on interactive sessions between these speakers, and lists research articles by project staff.
http://www.galanet.be/

Goethe-Institut

The website contains a linkpage focusing on Germany on the Internet.
This page includes links to online material for German teachers as well
as to chat rooms, fora and email contacts for learners of German.
http://www.goethe.de/lks/deindex.htm

Joint Information Systems Committee (JISC)

A UK-funded portal providing information in the innovative use of
information and communications technology to support what JISC calls
'e-learning', 'e-research' and 'e-resources' education and research
http://www.jisc.ac.uk/

Intute: arts and humanities

A merger of the former Humbul and Artifact sites, this is a free online
service providing access to the web resources for education and research,
selected and evaluated by a network of subject specialists. Keyword
searching and browsing are enabled.
http://www.intute.ac.uk/artsandhumanities/

Learning Technology Dissemination Initiative (LTDI)

Herriot-Watt University's resources site for teachers in higher education.
http://www.icbl.hw.ac.uk/ltdi/cookbook/

http://www.icbl.hw.ac.uk/ltdi/cookbook/
This LTDI link, based on a cookbook metaphor that may appeal or
irritate, leads to a simple and practical evaluation guide for higher
education practitioners carrying out small-scale projects.

National Foreign Language Resource Center

This Hawaii-based portal draws from foreign language teaching, applied linguistics and L2 acquisition to support projects that focus primarily on the less commonly taught languages of East Asia, Southeast Asia, and the Pacific. Free, but the site also gives access to materials that can be bought online. Covers both face-to-face and CMCL.
http://nflrc.hawaii.edu/index.cfm

Online Collaborative Learning in HE

Website run by Central Queensland University, with a range of resources on CSCL.
http://clp.cqu.edu.au

Online QDA (Qualitative Data Analysis)

Resource compiled by Gibbs, Fielding, Lewins and Taylor. Useful annotations and glossary. Many links to proprietary tools.
http://onlineqda.hud.ac.uk/index.php

Subject Centre for Languages, Linguistics and Area Studies

The UK Subject Centre's website lists events, links, papers, materials and other resources under the keyword 'Computer-Mediated Communication (CMC)' and 'Computer Assisted Language Learning (CALL)'.
http://www.lang.ltsn.ac.uk/resources/keywordresources.aspx?key wordid=386

Vance Stevens

Vance Stevens, an ESL teacher, has been active in the area of CALL and CMCL for a number of years. He developed the Webheads community of practice, and his website contains many relevant links for teachers and learners.
http://www.geocities.com/vance_stevens/vance.htm

Webheads

An online community of practice for teachers and learners of English. The focus is on Web 2.0 and computer-mediated communication. The site offers links, resources, online training, etc., and activities for students include 'Writing for Webheads' (e.g. synchronous text chat at Tapped In: http/tappedin.org/tappedin/). Teachers can join one of the 'Becoming a Webhead' workshops.
http://www.geocities.com/vance_stevens/papers/evonline2002/webheads. htm

18.4   Online bibliographies

Computer-Mediated Communication in Foreign Language
Education: An Annotated Bibliography

Coski and Kinginger's annotated bibliography. Valuable for the annotations, but the material listed is largely from early publications (up to 1999), so perhaps more of interest in a historical perspective.
http://nflrc.hawaii.edu/networks/NW03/default.html

18.5   Online books

New Perspectives on Call for Second Language Classrooms

Fotos and Browne's book is a print publication with an online version, available from some online university libraries. The publishers maintain a companion website for the book, offering links to further research and teaching resources (for full details, see Bibliography).
http://www.erlbaum.com/callforL2classrooms

ICT4LT

European-funded online resource on CALL for teachers in secondary and higher education. Although not published as an online book, the structure of the site allows it to be used as such. Module 2.3 has a useful section: *Towards Equal Discourse in Web-based Interaction.*
http://www.ict4lt.org/en/index.htm

Online tutoring e-Book

Free e-book, including chapters on learning, the tutor's role, online community building, assessment methods, evaluation, culture and ethics, institutional support and staff development, quality assurance and live links to many referenced articles. Chapters can be downloaded as pdf files.
http://otis.scotcit.ac.uk/onlinebook/

Theory and Practice of Online Learning

Anderson and Elloumi's 2004 free access book is published under a Creative Commons licence which allows you to read, print and share the contents freely (barring a few clearly spelt out restrictions). It is informed by the rich distance-teaching experience of Athabasca University (English-speaking Canada's open university), and covers four areas in CMC for learning (most of it applicable to languages): 'Role and Function of Theory in Online Education Development and Delivery'; 'Infrastructure

and Support for Content Development'; 'Design and Development of Online Courses'; 'Delivery, Quality Control, and Student Support of Online Courses'.
http://cde.athabascau.ca/online_book/index.html

18.6   Online journals

18.6.1   Journals with CMCL interest

ALSIC

Apprentissage des Langues et Systèmes d'Information et de Communication is the French journal for CMCL. It is online only and is free.
http://alsic.u-strasbg.fr or http://alsic.org

> ALSIC's resources page is newly edited with each issue of the journal. It hyperlinks professional associations, other journals, conferences, recent theses and many other items of interest.
> http://alsic.u-strasbg.fr/toiltheque/

Computer Assisted Language Learning

Intercontinental interdisciplinary journal covering all matters associated with the use of computers in language learning (L1 and L2), teaching and testing.
http://www.tandf.co.uk/journals/titles/09588221.asp

JALTCALL

All areas within the field of computer-assisted language learning, including teaching ideas and suggestions from teachers' personal experiences.
http://jaltcall.org/journal/

Language Learning and Technology

The most widely quoted international journal in CALL and CMCL.
http://llt.msu.edu/

> The journal's book reviews page give quick access to a very useful archive of reviews of major books in the field.
> http://llt.msu.edu/archives/bookstitle.html

PaCALL

The journal of the Pacific Association for Computer Assisted Language Learning promotes the use and professional support of CALL throughout the Pacific region, from North, East and Southeast Asia, to Oceania, and the coastal countries of the Americas.
http://www.paccall.org/journal/

ReCALL

The main journal in European CALL and CMCL.
http://journals.cambridge.org/action/displayJournal?jid=REC

System

Applications of technology and applied linguistics to foreign language teaching and learning, including EFL. Gives priority to theoretically based work.
http://www.elsevier.com/wps/find/journaldescription.cws_home/335/description#description

The CALICO Journal

Devoted to research and discussion on technology and language learning.
https://www.calico.org/p-5-CALICO%20Journal.html

18.6.2   Journals with CMC and educational technology interest

AJET

The *Australasian Journal of Educational Technology* publishes research and review articles in educational technology, instructional design, educational applications of computer technologies, educational telecommunications and related areas.
http://www.ascilite.org.au/ajet/ajet.html

ALT-J

The Association for Learning Technology's international tri-annual, peer-reviewed journal, devoted to research and good practice in the use of learning technologies within tertiary education.
http://www.alt.ac.uk/alt_j.html

British Journal of Educational Technology

Covers developments in educational technology world-wide in the fields of education, training and information technology.
http://www.blackwellpublishing.com/journal.asp?ref=0007-1013

Computers and Education

Welcomes articles on computer-mediated communication, cooperative and collaborative learning, country-specific developments, cross-cultural projects, distance education and telelearning, distributed learning environments.

http://www.elsevier.com/wps/find/journaldescription.cws_home/347/
description

International Journal of Computer-Supported Collaborative Learning

A peer-reviewed journal that promotes research on the theory and practice of computer-supported collaborative learning. A main focus is on how people learn. In the context of collaborative activity and how to design the technological settings for collaboration.
http://ijcscl.org/

Journal of Asynchronous Learning Networks

Aims to describe original work in asynchronous learning networks (ALN), including experimental results.
http://www.sloan-c.org/publications/jaln/

STICEF

Sciences et Technologies de l'Information et de la Communication pour l'Éducation et la Formation: a major French-language journal for computer environments and human learning (i.e. computer-assisted learning and CMC). Online only and free.
http://sticef.univ-lemans.fr/

The Journal of Computer-mediated Communication

Interdiciplinary journal devoted to CMC scholarship.
http://jcmc.indiana.edu/

The Journal of Interactive Online Learning

Interdisciplinary journal of theory, research, and practice in interactive online learning.
http://www.ncolr.org/jiol/

18.7   Online newsletters

THOT

Newsletter (in French) for distance education and training. Some content is free, the rest is pay-to-use, but it may be worthwhile subscribing to gain access to this rich resource, and have THOT alert you of new developments via email.
http://thot.cursus.edu/rubrique.asp?no=16402

18.8 Professional organisations

APACALL (Asia-Pacific Association for Computer-Assisted Language Learning)
http://www.apacall.org/

CALICO
Computer-Assisted Language Instruction Consortium
https://calico.org/

EALTA
European Association for Language Testing and Assessment
http://www.ealta.eu.org/

EUROCALL
European Association for Computer Assisted Language Learning
http://www.eurocall-languages.org/

IALLT
International Association for Language Learning Technology
http://iall.net/

PacCALL
Pacific Association for Computer-Assisted Language Learning
http://www.paccall.org/main/index.php

LET
The Japan Association for Language Education and Technology
http://www.j-let.org/

18.9 Tools and practical support: free

CLAPI

In French, this is the public access part of the research site 'Corpus de Langues Parlées en Interaction', run by researchers looking into interaction corpora. Audio and video extracts are available, both 'everyday' and educational. Useful for teaching speaking (possible use as examples of conversational stategies) or as a model for structuring one's own recorded corpora.
http://clapi.univ-lyon2.fr/

Free Computer Programs

Indexing and lexical frequency software from the French-Italian-Spanish Department of the University of Manitoba, Canada.
http://www.umanitoba.ca/fsi/fsicompt.htm

TESOL Quarterly

TESOL Quarterly guidelines for reporting quantitative research and three types of qualitative research: case studies, conversation analysis and (critical) ethnography.
http://www.tesol.org/pubs/author/serials/tqguides.html

University of Liège website

In French. A site on technology-based research methods and ethnomethodology. http://analyses.ishs.ulg.ac.be/logiciels/panorama.html

With an English page at:
http://analyses.ishs.ulg.ac.be/ethnomethodologie/links.html

University College London's Technology Research website

A series of simple templates for different types of dissemination articles drawn up by writers for the journal *ALT-J*, to help prepare for publishing work in the field of learning technologies, including case study, experiment, survey, etc.
http://www.ucl.ac.uk/calt/ltr/templates.html

Tropes

Designed for semantic classification and keyword extraction, Tropes software is very easy to use and can help with the content analysis of interaction logs, self-report accounts or open-ended questionnaire responses.
http://www.acetic.fr/ (French site, giving access to English, German, Spanish, Italian and Portuguese versions)

Utah University OpenCourseWare

An undergraduate course, 'Understanding Online Interaction', is available free of charge on this site. Easy, chatty style, it covers a range of systems, including games sites and includes assignments that repay if you spend a little time trying them out.
http://ocw.usu.edu/Instructional_Technology/Understanding_Online_Interaction

Web Style Guide

An online version of a widely used web style guide, useful for designing CMCL materials. It looks systematically at process, interface design, site and page design, typography, graphics and multimedia.
http://www.webstyleguide.com/index.html?/

Reinhard Donath: Englischunterricht in der Informationsgesellschaft
Useful tips for teachers
This website by an enthusiastic English teacher contains a range of information on online projects, books/articles, teaching practice and learning strategies. The site concentrates on English but also contains information on German as a foreign language and French.
http://www.englisch.schule.de/

Tandem

The tandem server at the Ruhr-Universität Bochum brings together learners of different languages by telephone, e-mail or other media. The service is free.
http://www.slf.ruhr-uni-bochum.de/

18.10   Tools and practical support: pay-to-use

ATLAS.ti

Tool for qualitative research able to handle multiple media. Developed in 1993 as part of a project at the Technical University Berlin. East Asian languages are supported. Claims to be easy to learn. Has an unlimited time download demo online.
http://www.atlasti.com/

HyperQual and SuperHyperQual

Two qualitative research tools. This is a very plain but clear website, with a pdf showing a selection of screens. The advantage of this software is that it allows researchers to maintain bibliographic databases for literature reviews linked to the data. No demo.
http://home.satx.rr.com/hyperqual/

Instat

Reading University's statistical services centre package.
http://www.rdg.ac.uk/ssc/software/instat/instat.html

LAMS

Learning Activity Management System is a visual authoring environment tool for designing technology-enabled course models, series of activities and lesson plans. LAMS also allows you to deliver courses to groups of students.
http://www.lamsinternational.com/

The Ethnograph

Tool for ethnographic research. The website has a page of very clear screen-shots, letting you see how easy the procedures are. The webpage on the demo version does not reveal the extent of the work that can be done, but asks you to fill in a form saying whether you wish to receive marketing information. Two chapters from its manual are available to download: 'Chapter 2: Overview' and 'Chapter 4: Quick Tour'.
http://www.qualisresearch.com/

N'VIVO

Possibly the best known of qualitative analysis packages. Handles vast quantities of data. Supports Japanese and Mandarin Chinese. Free 30 day download. A pdf explanatory brochure can be downloaded too but has no step-by-step procedure screenshots (hence better to download the free 30-page Guide to Getting Started).
http://www.qsr.com.au/products/productoverview/NVivo_7.htm

Statistics for the Terrified

As its name indicates, this site offers products with an appeal to those not experienced in quantitative analysis.
http://www.conceptstew.co.uk/PAGES/home.html

18.11   MOOs and virtual worlds

18.11.1   MOOs for language learning

Mundo hispano

http://www.umsl.edu/~moosproj/mundo.html
A text-based MOO for learning Spanish.

SchmoozeUniversity

http://schmooze.hunter.cuny.edu/
A text-based MOO for learning English.

MooFrançais

http://www.umsl.edu/~moosproj/moofrancais.html
A text-based MOO for learning French.

Diversity University

An English MOO. Research is being carried out here by many teachers and educators.
http://www.marshall.edu/commdis/moo/moo-connect.htm
http://www.diversityuniversity.com/

18.11.2 Graphical virtual worlds

Active Worlds

An avatar-based virtual world, which offers teachers and educators the Active Worlds Educational Universe (licence needed).
http://www.activeworlds.com/

Traveler

http://www.digitalspace.com/traveler/
Free virtual world.

Second Life

A three-dimensional virtual world with a growing number of members. It is becoming increasingly popular for educational use.

18.12 Video-streamed talks and other free educational sites

The Open University free Open Content pages (OpenLearn)

T180 Living with the Internet: learning Online

http://openlearn.open.ac.uk/mod/resource/view.php?id=79081

Open University learning and collaboration tools

http://labspace.open.ac.uk/course/view.php?id=2 and
http://labspace.open.ac.uk/mod/resource/view.php?id=54668

Video-streamed talks

SOLE (Spoken Online Learning Events), a two-day Seminar in June 2007 at the UK Open University, supported by the British Association for Applied Linguistics and Cambridge University Press. Webcast of Day 1 available at:
http:/stadium.open.ac.uk/stadia.preview.php?whichevent=994&s=29.

Webcast of Day 2 available at: http:/stadium.open.ac.uk/preview.php?whichevent=1012&s=29.

Video-streamed talks (in French) with an emphasis on language and technology, given at the Ecole Normale Supérieure des Lettres et Sciences Humaines (Lyons, France):

Online teaching: new professions (March 2003)
Collaborative learning online (March 2004)
Emotions and online interaction (March 2005)
All at:
http://ecole-ouverte.ens-lsh.fr/rubrique.php3?id_rubrique=115

18.13 Wikis

LanguageDaily

Currently in its infancy, this website aims at developing a wiki community for teaching yourself a foreign language and helping others along the way. It hopes to bring together language students, instructors and native speakers from all over the world.
http://languagedaily.com/home/index.php?title=Welcome%21

Bibliography

Abdullah, M. H. (1998) *Problem-Based Learning in Language Instruction: A Constructivist Model.* Eric Digest. ERIC Clearinghouse on Reading English and Communication, Bloomington, IN.

Aljaafreh, A. and J. P. Lantolf (1994) Negative Feedback as Regulation and Second Language Learning in the Zone of Proximal Development, *The Modern Language Journal* 78(4): 465–83.

Allum, P. (2002) CALL and the Classroom: The Case for Comparative Research, *ReCALL* 14(1): 146–66.

Allwright, D. (2005) Developing Principles for Practitioner Research: The Case of Exploratory Practice, *The Modern Language Journal* 89(3): 353–66.

Allwright, R. and J. Hanks (2007) *The Developing Language Learner*, Basingstoke: Palgrave Macmillan.

Anderson, T. and F. Elloumi (2004) *Theory and Practice of Online Learning*, Canada: Athabasca University (online publication): http://cde.athabascau.ca/ online_book/

Anderson-Mejias, P. (2006) A Case Study in Peer Evaluation for Second Language Teachers in Training, in T. Roberts (ed.), *Self, Peer and Group Assessment in E-Learning*, Hershey, PA: Information Science Publishing.

Atkinson, T. (2006) Computer Aided Assessment (CAA) and Language Learning. Module 4.1, in G. Davies (ed.), *ICT4LT, Information and Communications Technology for Language Teachers* (online resource): http://www.ict4lt.org/ en/en_mod1-5.htm

Baldry, A. and P. J. Thibault (2006) *Multimodal Transcription and Text Analysis: A Multimedia Toolkit and Coursebook with Associated Online Course.* London: Equinox.

Barbot, M.-J. and T. Lancien (eds) (2003) Médiation, Médiatisation et Apprentissages. *Notions en question.* Lyon: Editions de l'ENS.

Barr, D. (2004) *ICT – Integrating Computers in Teaching: Creating a Computer-based Language-learning Environment*, Oxford: Peter Lang.

Batardière, M-T. and C. Jeanneau (2006) *Quel est le bœuf?* Beefing up Language Classes with Collaborative Blogs. Paper presented at Integrating CALL into Study Programmes, EUROCALL Conference, Granada Spain.

Bax, S. (2003) CALL – Past, Present and Future, *System* 31(1): 13–28.

Bayer, A. S. (1990) *Collaborative Apprenticeship Learning: Language and Thinking across the Curriculum K-12.* Mountain View, CA: Mayfield Publishing.

Beatty, K. (2003) *Teaching and Researching: Computer-assisted Language Learning*, Harlow: Longman.

Beauvois, M. H. (1992) Computer-assisted Classroom Discussion in the Foreign Language Classroom: Conversations in Slow Motion, *Foreign Language Annals* 25(5): 455–64.

Beauvois, M. H. (1998) Conversations in Slow Motion: Computer-mediated Communication in the Foreign Language Classroom, *The Canadian Modern Language Review* 54 (2): 198–217.

Belz, J. A. (2001) Institutional and Individual Dimensions of Transatlantic Group Work in Network-based Language Teaching, *ReCALL* 13(2): 213–31.

Belz, J. A. (2002a) Second Language Play as a Representation of the Multicompetent Self in Foreign Language Study, *Journal for Language, Identity, and Education* 1: 13–39.

Belz, J. A. (2002b) Social Dimensions of Telecollaborative Foreign Language Study, *Language Learning and Technology* 6(1): 60–81.

Belz, J. A. (2003) Linguistic Perspectives on the Development of Intercultural Competence in Telecollaboration, *Language Learning and Technology* 7(2): 68–99.

Belz, J. A. (2006) At the Intersection of Telecollaboration, Learner Corpus Analysis, and L2 Pragmatics: Considerations for Language Program Direction, in J. A. Belz and S. L. Thorne (eds) *AAUSC 2005 – Internet-mediated Intercultural Foreign Language Education*, Boston, MA: Thomson Heinle and Heinle, 207–46.

Belz, J. A. (forthcoming) The Making and Masking of Linguistic Identity in Online Intercultural Relationship Building, in R. Goodfellow and M.-N. Lamy (eds), *Learning Cultures in Online Education*, London: Continuum Books.

Belz, J. A. and A. Müller-Hartmann (2003) Teachers as Intercultural Learners: Negotiating German–American Telecollaboration along the Institutional Fault Line, *The Modern Language Journal* 87(1): 71–89.

Belz, J. A. and S. L. Thorne (eds) (2006) *AAUSC 2005 – Internet-mediated Intercultural Foreign Language Education*, Boston, MA: Thomson Heinle and Heinle.

Benfield, G. (2000) Teaching on the Web: Exploring the Meanings of Silence (online publication): www.http://ultibase.rmit.edu.au/Articles/online/benfield1.htm

Bennett, S. and D. Marsh (2002) On Being an Online Tutor, *Innovations in Education and Teaching International* 39(1): 14–20.

Benson, P. (2001) *Teaching and Researching Autonomy in Language Learning*, Harlow: Longman.

Berge, Z. (2000) Components of the Online Classroom, in R. Weiss, D. S. Knowlton and B. W. Speck (eds), *Principles of Effective Teaching in the Online Classroom: New Directions for Teaching and Learning* 84: 23–28.

Betbeder, M.-L., C. Reffay and T. Chanier (2006) Environnement audio-graphique synchrone: recueil et transcription pour l'analyse des interactions multi-modales, in M. Sidir, E. Bruillard and G.-L. Baron (eds), *Proceedings of JOCAIR 2006, Premières journées communication et apprentissage instrumentés en réseau*, Amiens: Université de Picardie Jules Verne, 406–20.

Blake, R. (2000) Computer-mediated Communication: A Window on L2 Spanish Interlanguage, *Language Learning and Technology* 4(1): 120–36.

Block, D. (2003) *The Social Turn in Second Language Acquisition*, Edinburgh: Edinburgh University Press.

Blood, R. (2002) *We've Got Blog: How Weblogs are Changing Culture*, Cambridge, MA: Perseus Publishing, 7–16.

Bonk, C. and D. Cunningham (1998) Searching for Learner-Centered, Constructivist, and Sociocultural Components of Collaborative Educational Learning Tools, in C. Bonk and K. King (eds), *Electronic Collaborators: Learner-centered Technologies for Literacy, Apprenticeship and Discourse*, Mahwah, NJ: Lawrence Erlbaum Associates, 25–50.

Brown, E. (ed.) (2001) Mobile Learning Explorations at the Stanford Learning Lab, *Speaking of Computers* 55 (online publication): http://sll.stanford.edu/projects/tomprof/newtomprof/postings/289.html

Buckett, J., G. B. Stringer and N. K. J. Datta (1999) Life after ReLaTe: Internet Videoconferencing's Growing Pains, in K. Cameron (ed.), *CALL and the Learning Community*. Exeter: Elm Bank Publications: 31–8.

Buddyspace, described at http://buddyspace.sourceforge.net/ and pedagogical applications listed at http://www.aktors.org/technologies/buddyspace/

Bump, J. (1990) Radical Changes in Class Discussion Using Networked Computers, *Computers and the Humanities* 24: 44–65.

Burdeau, I. (1997) Virtual Classrooms, Virtual Schools: The New Possibilities, Paradigms and Pedagogy Created by Synchronous Computer-mediated Communication. Unpublished dissertation for the MA in Media Assisted Language Teaching, University of Brighton.

Burns, A. (1999) *Collaborative Action Research for English Language Teachers*, Cambridge: Cambridge University Press.

Burns, A. (2005) Action Research: An Evolving Paradigm? *Language Teaching* 38(2): 57–74.

Butler, J. (1990) *Gender Trouble: Feminism and the Subversion of Identity*, London: Routledge.

Butler, M. and S. Fawkes (1999) Videoconferencing for Language Learners, *Journal of the Association of Language Learning* 19: 46–9.

Bygate, M. and V. Samuda (2007) *Tasks in Second Language Learning*. Basingstoke: Palgrave Macmillan.

Cameron, D. (2002) *Working with Spoken Discourse*. London: Sage.

Cameron, K. (1997) Editorial, *Computer Assisted Language Learning* 10(5): 409–10.

Campbell, A. P. (2003) Weblogs for Use with ESL Classes, *The Internet TESL Journal* 9(2) (online Publication): http://iteslj.org/Techniques/Campbell-Weblogs. html

Canagarajah, S. (2002) *Critical Academic Writing and Multilingual Students*, Ann Arbor, MI: University of Michigan Press.

Candlin, C. and F. Byrnes (1995) Designing for Open Language Learning: Teaching Roles and Learning Strategies, in S. Gollin (ed.), *Language in Distance Education: How Far Can We Go?* Proceedings of the National Centre for English Language Teaching and Research Mini-conference, Perth 1994, Sydney: NCELTR.

Candlin, S. and N. C. Candlin (2007) Nursing through Time and Space: Some Challenges to the Construct of Community of Practice, in R. Iedema (ed.), *Discourse of Hospital Communication: Tracing Complexities in Contemporary Health Organisations*, Basingstoke: Palgrave Macmillan.

Candlin, C. N. and D. Murphy (1987) *Language Learning Tasks*, Englewood Cliffs, NJ: Prentice-Hall International.

Carr, W. and S. Kemmis (1986) *Becoming Critical: Education, Knowledge and Action Research*, Lewes: Falmer Press.

Chanier, T. and A. Vetter (2006) Multimodalité et expression en langue étrangère dans une plateforme audio-synchrone, *Apprentissage des Langues et Systèmes d'Information et de Communication* 9(1): 61–70.

Chapelle, C. A. (1997) CALL in the Year 2000: Still in Search of Research Paradigms? *Language Learning and Technology* 1(1): 19–43.

Chapelle, C. A. (2000) Is Network-based Learning CALL?, in M. Warschauer and R. Kern (eds), *Network-Based Language Teaching: Concepts and practice*, Cambridge: Cambridge University Press.

Chapelle, C. A. (2001) *Computer Applications in Second Language Acquisition: Foundations for Teaching, Testing and Research*, Cambridge: Cambridge University Press.

Chapelle, C. A. (2003) *English Language Learning and Technology*, Amsterdam and Philadelphia, PA: John Benjamins.

Chapelle, C. A. and D. Douglas (2006) *Assessing Language through Computer Technology*, Cambridge Language Assessment, Cambridge: Cambridge University Press.

Chaptal, A. (2003) Réflexions sur les technologies éducatives et les évolutions des usages: le dilemme constructiviste, *Distances et Savoirs* 1(1): 122–47.

Chinnery, G. M. (2006) Going to the MALL: Mobile-assisted Language Learning, *Language Learning and Technology* 10(1): 9–16.

Chiu, C.Y. and S. Savignon (2006) Writing to Mean: Computer-mediated Feedback in Online Tutoring of Multidraft Compositions, *The CALICO Journal* 24(1): 97–114.

Chun, D. M. (1994) Using Computer Networking to Facilitate the Acquisition of Interactive Competence, *System* 22(1): 17–31.

Chun, D.M. (2007) Come Ride the Wave: But Where is it Taking Us? *The CALICO Journal* 24(2): 239–52.

Clerehugh, Y. J. (2002) Information and Communication Technology as a Motivator for Disaffected Pupils, *Pedagogy, Culture and Society* 10(2): 209–22.

Colpaert, J. (2004) From Courseware to Coursewear? *Computer-Assisted Language Learning*, 17(3/4): 261–6.

Coniam, D. (2006) Evaluating Computer-based and Paper-based Versions of an English-language Listening Test, *ReCALL* 18(2): 193–211.

Conole, G. and M. Dyke (2004a) What are the Affordances of Information and Communication Technologies? *ALT-J, Research in Learning Technology* 12(2): 113–24.

Conole, G. and M. Dyke (2004b) Understanding and Using Technological Affordances: A Response to Boyle and Cook, *ALT-J, Research in Learning Technology* 12(3): 302–8.

Cook, G. (2003) Various Shades of Grey: The Losses and Gains of Contemporary Multimodality. Paper presented at BAAL/CUP seminar on Multimodality and Applied Linguistics, University of Reading 18–19 July.

Coverdale-Jones, T. (1998) Does Computer-mediated Conferencing Really Have a Reduced Social Dimension? *ReCALL* 10(1): 46–52.

Csikszentmihalyi, M. (1990) *Flow: The Psychology of Optimal Experience*, New York: Harper & Row.

Cziko, G. A. and S. Park (2003) Internet Audio Communication for Second Language Learning: A Comparative Review of Six Programs, *Language Learning and Technology* 7(1): 15–27.

Dam, L. (1995) *Learner Autonomy 3: From Theory to Classroom Practice*, Dublin: Authentik.

Daradoumis, T., F. Xhafa and J. Pérez (2006) A Framework for Assessing Self, Peer, and Group Performance in E-Learning, in T. Roberts (ed.), *Self, Peer and Group Assessment in E-Learning*, Hershey, PA: Information Science Publishing.

Davies, A. (2003) *The Native Speaker: Myth and Reality*, Clevedon: Multilingual Matters.

Davis, B. and R. Thiede (2000) Writing into Change: Style Shifting in Asynchronous Electronic Discourse, in M. Warschauer and R. Kern (eds), *Network-based Language Teaching: Concepts and Practice*, Cambridge: Cambridge University Press.

de Bot, K., W. Lowie and M. Verspoor (2005) *Second Language Acquisition: An Advanced Resource Book*, London: Routledge.

De Laat, M. (2006) *Networked Learning*, Politieacademie: Universiteit Utrecht (online publication): http://www.e-learning.nl/files/ dissertatie%20maarten. pdf

Debski, R. (1997) Support of Creativity and Collaboration in the Language Classroom: A New Role for Technology, in R. Debski, J. Gassin and M. Smith (eds), *Language Learning through Social Computing*, Melbourne: University of Melbourne, 39–65.

Debski, R. (2003) Analysis of Research in CALL (1980–2000) with a Reflection on CALL as an Academic Discipline, *ReCALL* 15(2): 177–88.

Debski, R. and M. Levy (1999) Introduction, in R. Debski and M. Levy (eds), *WORLDCALL: Global Perspectives on Computer-assisted Language Learning*. Lisse: Swets and Zeitlinger, 8–10.

Dias, J. (1998) The Teacher as Chameleon: Computer-mediated Communication and Role Transformation, in P. Lewis (ed.), *Teachers, Learners, and Computers: Exploring Relationships in CALL*, Nagoya: The Japanese Association for Language Teaching, 17–26.

Dias, J. (2002) Cell Phones in the Classroom: Boon or Bane? Part 2, *C@lling Japan: The Newsletter of the JALT-CALL Special Interest Group*, 10(3): 8–13 (online publication): http://jaltcall.org/cjo/10_3.pdf

Doering, A. and R. Beach (2002) Preservice English Teachers Acquiring Literacy Practices through Technology Tools, *Language Learning and Technology* 6(3): 127–46.

Donath, R. (2002) E-Mail-Projekte im Englischunterricht: Tipps für die Unterrichtspraxis. *Englischunterricht in der Informationsgesellschaft* (online publication): http://www.englisch.schule.de/tipps_neu.htm#zehn

Donato, R. and D. McCormick (1994) A Sociocultural Perspective on Language Learning Strategies: The Role of Mediation, *The Modern Language Journal* 78(4): 453–64.

Dörnyei, Z. (1994) Motivation and Motivating in the Foreign Language Classroom, *Modern Language Journal* 78: 273–84.

Dörnyei, Z. (2001a) *Teaching and Researching Motivation*, Harlow: Longman.

Dörnyei, Z. (2001b) *Motivation Strategies in the Language Classroom*, Cambridge: Cambridge University Press.

Dörnyei, Z. (2003) *Questionnaires in Second Language Research: Construction, Administration and Processing*, Mahwah, NJ: Lawrence Erlbaum Associates.

Doughty, C. and M. H. Long (2003) Optimal Psycholinguistic Environments for Distance Foreign Language Learning, *Language Learning and Technology* 7(3): 50–80.

Dresner, E. (2006) Textual Multitasking in CMC: Implications and Applications, *Interface: The Journal of Education, Community and Values* (online publication): http://bcis.pacificu.edu/journal/2006/01/dresner.php

Egbert, J. L. and G. M. Petrie (eds) (2005) *CALL Research Perspectives*, Mahwah, NJ: Lawrence Erlbaum Associates.

Ellis, R. (2000) Task-based Research and Language Pedagogy, *Language Teaching Research* 4(3): 193–220.

Ellis, R. (2003) *Task-based Language Learning and Teaching*, Oxford: Oxford University Press.

Embedding Learning Technologies (online resource): www.elt.ac.uk/ materials. htm

Engler, L.-R. (2001) Deutsch lernen über das Internet: Die Möglichkeiten eines didaktischen Chats, *Linguistik online* 9(2) (online publication): http://www. linguistik-online.de/9_01/Engler.html

Engstrom, M. E. and D. Jewett (2005) Collaborative Learning the Wiki Way, *TechTrends: Linking Research and Practice to Improve Learning* 49(6): 12–15.

Erben, T. (1999) Constructing Learning in a Virtual Immersion Bath: LOTE Teacher Education through Audiographics, in R. Debski and M. Levy (eds), *WORLDCALL: Global Perspectives on Computer-assisted Language Learning*, Lisse: Swets and Zeitlinger, 229–48.

Erlich, Z., I. Erlich-Philip and J. Gal-Ezer (2005) Skills Required for Participating in CMC Courses: An Empirical Study, *Computers and Education* 44: 477–87.

Fanderclai, T. L. (1995) MUDs in Education: New Environments, New Pedagogies. *Computer-Mediated Communication Magazine* 2(1) (online publication): http:// www.ibiblio.org/cmc/mag/1995/jan/fanderclai.html

Felix, U. (2005) What Do Meta-analyses Tell Us about CALL Effectiveness? *ReCALL* 17(2): 269–88.

Finneran, C. M. and P. Zhang (2003) A Person–Artefact–Task (PAT) Model of Flow Antecedents in Computer-mediated Environments, *International Journal of Human–Computer Studies* 50: 475–96.

Fitzpatrick, A. and G. Davies (eds) (2003) The Impact of Information and Communications Technologies on the Teaching of Foreign Languages and on the Role of Teachers of Foreign Languages: A Report Commissioned by the Directorate General of Education and Culture (online publication): http://ec.europa.eu/education/policies/lang/doc/ict.pdf

Fotos, S. and C. M. Browne (2004) The Development of CALL and Current Options, in S. Fotos and C. M. Browne (eds), *New Perspectives on Call for Second Language Classrooms*, Mahwah, NJ: Lawrence Erlbaum Associates, 3–14.

Fulcher, G. and F. Davidson (2007) *Language Testing and Assessment*. New York: Routledge.

Furstenberg, G. (1997) Teaching with Technology: What is at Stake? *Association of Departments of Foreign Languages Bulletin* 28 (3): 21–5.

Furstenberg, G., S. Levet, K. English and K. Maillet (2001) Giving a Virtual Voice to the Silent Language of Culture: The Cultura Project, *Language Learning and Technology* 5(1): 55–102.

Garcia, A. C. and J. B. Jacobs (1999) The Eyes of the Beholder: Understanding the Turn-taking System in Quasi-synchronous Computer-mediated Communication, *Research on Language and Social Interaction* 32(4): 337–67.

Gardner, R. C., P. Tremblay and A.-M. Masgoret (1997) Towards a Full Model of Second Language Learning: An Empirical Investigation, *The Modern Language Journal* 81 (3), 344–62.

Gardner, R. and J. Wagner (2004) *Second Language Conversations*, London: Continuum Books.

Gass, S. (1997) *Input, Interaction, and the Second Language Learner*, Mahwah, NJ: Lawrence Erlbaum Associates.

Gass, S. and A. Mackey (2000) *Stimulated Recall Methodology in Second Language Research*, Mahwah, NJ: Lawrence Erlbaum Associates.

Gass, S. M., A. Mackey and T. Pica (1998) The Role of Input and Interaction in Second Language Acquisition, *Modern Language Journal* 82: 299–307.

Gass, S. and E. M. Varonis (1994) Input, Interaction, and Second Language Production, *Studies in Second Language Acquisition* 16(3): 283–302.

Gee, J. P. (2000) New People in New Worlds: Networks, the New Capitalism and Schools, in B. Cope and M. Kalantzis (eds), *Multiliteracies: Literacy Learning and the Design of Social Futures*, London: Routledge, 43–68.

Gibbs, G. R., N. Fielding, A. Lewins and C. Taylor, *Online QDA* (online resource): http://onlineqda.hud.ac.uk/index.php

Gibson, J. J. (1979) *The Ecological Approach to Visual Perceptions*, Boston, MA: Houghton Mifflin.

Godwin-Jones, R. (1999) Mobile Computing and Language Learning, *Language Learning and Technology* 2(2): 7–11.

Godwin-Jones, R. (2001) Emerging Technologies: Language-testing Tools and Technologies, *Language Learning and Technology* 5(2): 8–12.

Godwin-Jones, R. (2003) Emerging Technologies Blogs and Wikis: Environments for On-line Collaboration, *Language Learning and Technology* 7(2): 12–16.

Godwin-Jones, R. (2005) Messaging, Gaming, Peer-to-Peer Sharing: Language Learning Strategies and Tools for the Millennial Generation, *Language Learning and Technology* 9 (1): 17–22.

Goffman, E. (1967) *Interaction Rituals: Essays on Face-to-Face Behavior*, New York: Random House.

Goodfellow, R. (2001) Credit Where it's Due: Assessing Students' Contributions to Collaborative Online Learning, in D. Murphy, R. Walker and G. Webb (eds), *Online Learning and Teaching with Technology: Case Studies, Experience and Practice*, London: Kogan Page, 73–80.

Goodfellow, R. (2004) Online Literacies and Learning: Operational, Cultural and Critical Dimensions, *Language and Education*, 18(5): 379–99.

Goodfellow, R. and A. Hewling (2005) Re-conceptualising Culture in Virtual Learning Environments: From an 'Essentialist' to a 'Negotiated' Perspective, *e-Learning* 2(4): 356–68.

Goodfellow, R., I. Jefferys, T. Miles and T. Shirra (1996) Face-to-Face Language Learning at a Distance? A Study of a Videoconference Try-out, *ReCALL* 8(2): 5–16.

Goodfellow, R., M.-N. Lamy and G. Jones (2002) Assessing Learners' Writing Using Lexical Frequency, *ReCALL* 14(1): 133–45.

Goodyear, P., S. Banks, V. Hodgson and D. McConnell (2004) Research on Networked Learning: An Overview, in P. Goodyear, S. Banks, V. Hodgson and D. McConnell (eds), *Advances in Research on Networked Learning*, Boston, MA: Kluwer, 1–9.

Goodyear, P., G. Salmon, J. M. Spector, C. Steeples and S. Tickner (2001) Competences of Online Teaching: A Special Report, *Educational Technology Research and Development* 49(1): 65–72.

Grabe, W. and F. L. Stoller (2002) *Teaching and Researching Reading*. Harlow: Longman.

Gregersen, T. and E. K. Horwitz (2002) Language Learning and Perfectionism: Anxious and Non-anxious Language Learners' Reactions to Their Own Oral Performance, *The Modern Language Journal* 86(4): 562–70.

Gunawardena, C. N. (1995) Social Presence: Theory and Implications for Interaction and Collaborative Learning in Computer Conferences, *International Journal of Educational Telecommunications* 1(2/3): 147–66.

Hafner, C. A. (2006) Understanding Learner Behaviour in an Online Concordancing Environment. In J. Colpaert, W. Decoo, S. Van Beuren and A. Godfroid (eds), *CALL and Monitoring the Learner, Proceedings of the 12th International CALL Conference*, Antwerp: University of Antwerp, August, 96–9.

Hafner, C. (2007) Personal communication.

Halliday, M. A. K. (1993) Towards a Language-Based Theory of Learning, *Linguistics and Education* 5(2): 93–116.

Halliday, M. A. K. and R. Hasan (1976) *Cohesion in English*, Harlow: Longman.

Hampel, R. (2006) Rethinking Task Design for the Digital Age: A Framework for Language Teaching and Learning in a Synchronous Online Environment, *ReCALL* 18(1): 105–21.

Hampel, R., U. Felix, M. Hauck and J. Coleman (2005) Complexities of Learning and Teaching Languages in a Real-time Audiographic Environment, *German as a Foreign Language* 3, http://www.gfl-journal.de/3-2005/hampel_ felix_hauck_ coleman.pdf

Hampel, R. and M. Hauck (2004) Towards an Effective Use of Audio Conferencing in Distance Language Courses, *Language Learning and Technology* 8(1): 66–82.

Hampel, R. and U. Stickler (2005) New Skills for New Classrooms: Training Tutors to Teach Languages Online, *Computer Assisted Language Learning* 18(4): 311–26.

Hanna, B. E. and J. de Nooy (2003) A Funny Thing Happened on the Way to the Forum: Electronic Discussion and Foreign Language Learning, *Language Learning and Technology* 7(1): 71–85.

Hara, N. and R. Kling (1999) Students' Frustrations with a Web-based Distance Education Course. *First Monday* 4(12) (online publication): http://www. firstmonday.org/issues/issue4_12/hara/

Harasim, L. (1990) Online Education: An Environment for Collaboration and Intellectual Amplification, in L. Harasim (ed.), *Online Education: Perspectives on a New Environment*, New York: Praeger, 39–64.

Harrington, M. and M. Levy (2001) CALL Begins with a C: Interaction in Computer-mediated Language Learning, *System* 29: 15–26.

Hassan, X., D. Hauger, G. Nye and P. Smith (2005) The Use and Effectiveness of Synchronous Audiographic Conferencing in Modern Language Teaching and Learning (Online Language Tuition): A Systematic Review of Available Research, in *Research Evidence in Education Library*, London: EPPI-Centre, Social Science Research Unit, Institute of Education, University of London.

Hauck, M. and S. Hurd (2005) Exploring the Link between Language Anxiety and Learner Self-management in Open Language Learning Contexts, *European Journal of Open, Distance and E-Learning* 2005(2) (online publication): http://oro.open.ac.uk/3542/

Hauck, M. and U. Stickler (eds) (2006) What Does it Take to Teach Online? Towards a Pedagogy for Online Language Teaching and Learning, special issue of *The CALICO Journal* 23(3).

Heift, T. (2001) Error-specific and Individualized Feedback in a Web-based Language Tutoring System: Do They Read it? *ReCALL* 13(1): 99–109.

Heins, B. (2005) Personal communication.

Henri, F. and Lundgren-Cayrol, K. (1997) *Apprentissage collaboratif à distance, téléconférence et télédiscussion*. Internal Report 3 (version 1.7). Montreal: LICEF (online publication): http://www.licef.teluq.uquebec.ca/Bac/fiches/ f48.htm

Herring, S. (2004) Computer-mediated Discourse Analysis: An Approach to Researching Online Behavior, in S. A. Barab, R. Kling and J. H. Gray (eds), *Designing for Virtual Communities in the Service of Learning*, New York: Cambridge: Cambridge University Press: 338–76.

Hewson, C. (2003) Conducting Research on the Net, *The Psychologist* 6(6): 290–93.

Hoffman, R. (1993) The Distance Brings Us Closer: Electronic Mail, ESL Learner Writers, and Teachers, in G. Davies and B. Samways (eds), *Teleteaching '93*, Proceedings of the IFIP TC3, 3rd Teleteaching Conference, Trondheim 1993, Amsterdam: North Holland, 391–9.

Horwitz, E. K. (2000) 'It ain't over 'til it's over': On Foreign Language Anxiety, First Language Deficits and the Confounding of Variables, *The Modern Language Journal* 84: 256–59.

Hubbard, P. (2004) Learner Training for Effective Use of CALL, in S. Fotos and C. M. Browne (eds), *New Perspectives on Call for Second Language Classrooms*, Mahwah, NJ: Lawrence Erlbaum Associates, 45–67.

Hubbard, P. (2005) A Review of Characteristics in CALL Research, *Computer Assisted Language Learning* 18(5): 351–68.

Humlab Blog, http://blog.umlab.umu.se/

Hunt, M. (2001) Principles and Theoretical Approaches in Assessment, in L. Arthur and S. Hurd (eds), *Supporting Lifelong Language Learning: Theoretical and Practical Approaches*, London: Centre for Information for Language Teaching and Research, 152–64.

Hutchby, I. (2001) *Conversation and Technology: From the Telephone to the Internet*, Cambridge: Polity Press.

Jazwinski, C. H. (2001) Gender Identities on the World Wide Web, in C. R. Wolfe (ed.), *Learning and Teaching on the World Wide Web*, San Diego: Academic Press.

Johnson, K. (2003) *Designing Language Teaching Tasks*, Basingstoke: Palgrave Macmillan.

Jones, R. H. (2004) The Problem of Context in Computer-Mediated Communication, in P. Levine and R. Scollon (eds), *Discourse and Technology: Multimodal Discourse Analysis*, Washington, DC: Georgetown University Press, 20–33.

Jung, U. (2005) CALL – Past, Present and Future: A Bibliometric Approach, *ReCALL* 17(1): 4–17.

Kelm, O. R. (1998) The Use of Electronic Mail in Foreign Language Classes, in J. Swaffar, S. Romano, P. Markley and K. Arens (eds), *Language Learning Online: Theory and Practice in the ESL and L2 Computer Classroom*, Austin, TX: Labyrinth Publications, 141–53.

Kemmis, S. and R. McTaggart (1988) *The Action Research Planner*, 3rd edition, Victoria: Deakin University.

Kern, R. G. (1995) Restructuring Classroom Interaction with Networked Computers: Effects on Quantity and Characteristics of Language Production, *The Modern Language Journal* 79(4): 457–76.

Kern, R. G. (2006) La Communication médiatisée par ordinateur en langues: recherches et applications récentes aux USA, in F. Mangenot and C. Dejean-Thircuir (eds), *Les Echanges en ligne dans l'apprentissage et la formation, Le français dans le monde, Recherches et Applications* 40: 17–29.

Kern, R. G., Ware P. and M. Warschauer (2004) Crossing Frontiers: New Directions in Online Pedagogy and Research, *Annual Review of Applied Linguistics* 24: 243–60.

Kiernan, J. P. and K. Aizawa (2004) Are Cell Phones Useful Language Learning Tools? *ReCALL* 16(1): 71–84.

Kinginger, C. (1998) Videoconferencing as Access to Spoken French, *The Modern Language Journal* 82(4): 502–13.

Kinginger, C., A. Gourves-Hayward and V. Simpson (1999) A Telecollaborative Course on French–American Intercultural Communication, *French Review* 72(5): 853–66.

Kirkwood, A. and L. Price (2005) Learners and Learning in the Twenty-first Century: What Do We Know about Students' Attitudes towards and Experiences of Information and Communication Technologies that will Help Us Design Courses? *Studies in Higher Education* 30(3): 257–74.

Kötter, M. (2001) Developing Distance Language Learners' Interactive Competence: Can Synchronous Audio Do the Trick? *International Journal of Educational Telecommunications* 7(4): 327–53.

Kötter, M. (2003) Negotiation of Meaning and Codeswitching in Online Tandems, *Language Learning and Technology* 7(2): 145–72.

Kramsch, C. (1986) From Language Proficiency to Interactional Competence, *Modern Language Journal* 70: 366–72.

Kramsch, C. (ed.) (2002) *Language Acquisition and Language Socialization: Ecological Perspectives*, London: Continuum Books.

Kramsch, C. and S. L. Thorne (2001) Foreign Language Learning as Global Communicative Practice, in D. Block and D. Cameron (eds), *Globalization and Language Teaching*, London: Routledge: 83–100.

Krashen, S. (1981) *Second Language Acquisition and Second Language Learning*, Oxford: Pergamon Press.

Krashen, S. (1985) *The Input Hypothesis: Issues and Implications*, Harlow: Longman.

Kress, G. (2000) Design and Transformation: New Theories of Meaning, in B. Cope and M. Kalantzis for the New London Group (eds), *Multiliteracies: Literacy Learning and the Design of Social Futures*, London: Routledge, 153–61.

Kress, G. (2003) *Literacy in the New Media Age*, London: Routledge.

Kress, G., C. Jewitt, J. Ogborn and C. Tsatsarelis (2001) *Multimodal Teaching and Learning: The Rhetorics of the Science Classroom*, London: Continuum Books.

Kress, G. and T. van Leeuwen (2001) *Multimodal Discourse: The Modes and Media of Contemporary Communication*, London: Arnold.

Kukulska-Hulme, A. M. (2007) Mobile Language Learning Now and in the Future, in P. Svensson (ed.), *Från vision till praktik: Språkutbildning och Informationsteknik* [*From Vision to Practice: Language Learning and IT*], Härnösand: Myndigheten för nätverk och samarbete inom högre utbildning., Swedish Net University, 295–310.

Kukulska-Hulme A. M., D. Evans and J.Traxler (2005) Landscape Study in Wireless and Mobile Learning in the post-16 sector (online publication): http://www.jisc.ac.uk/uploaded_documents/SUMMARY%20FINAL%202005.doc

Kukulska-Hulme, A. and L. Shield (2004) Usability and Pedagogical Design: Are Language Learning Websites Special? *Proceedings of the ED-MEDIA'04 World Conference on Educational Multimedia, Hypermedia and Telecommunications*, Association for the Advancement of Computing in Education Digital Library

2004(1): 4235–42 (online publication): http://www.aace.org/DL/index.cfm? fuseaction=ViewPaper&id=16072

Lam, P. and C. McNaught (2006) Evaluating Designs for Web-assisted Peer and Group Assessment, in T. Roberts (ed.). *Self, Peer and Group Assessment in E-Learning*, Hershey, PA: Information Science Publishing.

Lamy M.-N. (2004) Oral Conversations Online: Redefining Oral Competence in Synchronous Environments, *ReCALL* 16(2): 520–38.

Lamy, M.-N. and R. Goodfellow (1999) Reflective Conversations in the Virtual Language Classroom, *Language Learning and Technology* 2(2): 43–61.

Lamy, M.-N. and X. Hassan (2003) What Influences Reflective Interaction in Distance Peer Learning? Evidence from Four Long-term Online Learners of French, *Open Learning* 18(1): 39–59.

Lamy, M.-N and H. J. Klarskov (2000) Using Concordance Programs in the Modern Foreign Languages Classroom, in G. Davies (ed.), *ICT 4 LT: Information and Communication Technology for Language Teachers* (online resource): http://www.ict4lt.org/en/index.htm

Lan, Y. Y. Sung and K. Chan (2006) What Makes Collaborative Early EFL Reading Effective? A Mobile Dynamic Peer-assisted Learning System. Paper presented at the IADIS International Conference on Mobile Learning, San Sebastian (online publication): http://www.ntnu.edu.tw/acad/docmeet/95/a1/a103-1.doc

Lankshear, C., J. P. Gee, M. Knobel and C. Searle (1997) *Changing Literacies*, Buckingham and Philadelphia: Open University Press.

Lankshear, C. and M. Knobel (2003a) *New Literacies: Changing Knowledge and Classroom Learning*, Buckingham and Philadelphia: Open University Press.

Lankshear, C. and M. Knobel (2003b) Do-it-Yourself Broadcasting: Writing Weblogs in a Knowledge Society. Paper presented at the American Educational Research Association Conference, Chicago (online publication): http://www. geocities.com/c.lankshear/blog2003.html?200627

Lankshear, C. and M. Knobel (2006) Blogging as Participation: The Active Sociality of a New Literacy. Paper presented at the American Educational Research Association, San Francisco (online publication): http://www. geocities.com/c.lankshear/bloggingparticipation.pdf

Lantolf, J. P. (2000) Second Language Learning as a Mediated Process, *Language Teaching* 33(2): 79–96.

Lantolf, J. and A. Aljaafreh (1995) Second Language Learning in the Zone of Proximal Development: A Revolutionary Experience, *International Journal of Educational Research* 23: 619–32.

Lantolf, J. and G. Appel (1994) *Vygotskian Approaches to Second Language Research*, Norwood, NJ: Ablex.

Lantolf, J. and A. Pavlenko (1996) Sociocultural Theory and Second Language Acquisition, *Annual Review of Applied Linguistics* 15: 108–24.

Lantolf, J. P. and S. L. Thorne (2006) *Sociocultural Theory and the Genesis of Second Language Development*, London: Oxford University Press.

Laurier, M. (2003) Can Computerized Testing be Authentic? *ReCALL* 12 (1): 93–104.

Laurillard, D., M. Stratfold, R. Luckin, L. Plowman and L. Taylor (2000) Affordances for Learning in a Non-linear Narrative Medium, *Journal of Interactive Media in Education* 2: 1–19.

Lave, J. (1991) Situating Learning in Communities of Practice, in L. B. Resnick, J. M. Levine and S. D. Teasley (eds), *Perspectives on Socially Shared Cognition*, Washington, DC: American Psychological Association, 63–82.

Lave, J. and E. Wenger (1991) *Situated Learning: Legitimate Peripheral Participation*, Cambridge: Cambridge University Press.

Lea, M. (2005) 'Communities of Practice' in Higher Education: Useful Heuristic or Educational Model? in D. Barton and K. Tusting (eds), *Beyond Communities of Practice: Language Power and Social Context*, Cambridge: Cambridge University Press.

Lecourt, D. (1999) The Ideological Consequences of Technology and Education: The Case for Critical Pedagogy, in M. Selinger and J. Pearson (eds), *Telematics in Education: Trends and Issues*, Amsterdam: Pergamon, 51–75.

Lee, L. (2002a) Synchronous Online Exchanges: A Study of Modification Devices on Non-native Discourse, *System* 30: 275–88.

Lee, L. (2002b) Enhancing Learners' Communication Skills through Synchronous Electronic Interaction and Task-Based Instruction, *Foreign Language Annals* 35(1): 16–23.

Leloup, J. W. and R. Ponterio (2003) Second Language Acquisition and Technology: A Review of the Research, *ERIC Digest* (EDO-FL-03-11): 1–2.

Lemke, J. (2002) Language Development and Identity: Multiple Timescales in the Social Ecology of Learning, in C. Kramsch (ed.), *Language Acquisition and Language Socialization*. London and New York, Continuum: 1–30.

Lemke, J. (2006) Towards Critical Multimedia Literacy: Technology, Research, and Politics, in M. McKenna, L. Labbo, D. Reinking and R. Kieffer (eds), *International Handbook of Literacy and Technology* II. Mahwah, NJ: Lawrence Erlbaum Associates, 3–14.

Leontiev, A. (1981) *Psychology and the Language-learning Process*, Oxford: Pergamon Press.

Lessig, L. (2004) *Free Culture: How Big Media Uses Technology and the Law to Lock Down Culture and Control Creativity*, New York: Penguin Books.

Levy, M. (1998) Two Conceptions of Learning and Their Implication for CALL at the Tertiary Level, *ReCALL* 10(1): 86–94.

Levy, M. (2000) Scope, Goals and Methods in CALL Research: Questions of Coherence and Autonomy, *ReCALL* 12(2): 170–95.

Levy, M. (2007) Culture, Culture Learning and New Technologies: Towards a Pedagogical Framework, *Language Learning and Technology* 11(2): 104–127.

Levy, M. and P. Hubbard (2005) Why call CALL 'CALL'? *Computer Assisted Language Learning* 18(3): 143–9.

Levy, M. and G. Stockwell (2006) *CALL Dimensions: Options and Issues in Computer-assisted Language Learning*. Mahwah, NJ: Lawrence Erlbaum Associates.

Lewin, K. (1948) *Resolving Social Conflicts*, New York: Harper & Row.

Lewins, A. and C. Silver (2007) *Using Software in Qualitative Research: A Step-by-Step Guide*, London: Sage.

Lewis T. W. (2006) When Teaching is Learning: A Personal Account of Learning to Teach Online, *The CALICO Journal* 23(3): 581–600.

Little, D. (1991) *Learner Autonomy 1: Definitions, Issues and Problems*, Dublin: Authentik.

Liu, M. Z. Moore, L. Graham and S. Lee (2002) A Look at the Research on Computer-based Technology Use in Second-language Learning: A Review of the

Literature from 1990–2000, *Journal of Research on Technology in Education* 34(3): 250–73.

Long, M. H. (1983) Linguistic and Conversational Adjustments to Non-Native Speakers, *Studies in Second Language Acquisition* 5(2): 177–93.

Long, M. H. and P. Robinson (1998) Focus on Form: Theory, Research, and Practice, in C. Doughty and J. Williams (eds), *Focus on Form in Classroom Second Language Acquisition*, Cambridge: Cambridge University Press, 15–41.

Luke, C. (2000) Cyber-schooling and Technological Change: Multiliteracies for New Times, in B. Cope and M. Kalantzis (eds), *Multiliteracies: Literacy Learning and the Design of Social Futures*. London: Routledge, 69–91.

Lund, A. (2006) The Multiple Contexts of Online Language Teaching, *Language Teaching Research* 10(2): 181–204.

Lund, A. and O. Smørdal, (2006) Is There a Space for the Teacher in a Wiki?, *Proceedings of the 2006 International Symposium on Wikis (WikiSym '06)*, Odense, Denmark: ACM Press, 37–46 (online publication): http://www.wikisym.org/ws2006/proceedings/p37.pdf

Macaro, E. (1997) *Target Language, Collaborative Learning and Autonomy*, Clevedon: Multilingual Matters.

Macdonald, J. (2003) Assessing Online Collaborative Learning: Process and Product, *Computers and Education* 40(4): 377–91.

Macdonald, J. (2004) Developing Competent e-Learners: The Role of Assessment, *Assessment and Evaluation in Higher Education* 29(2): 215–26.

Mangenot, F. (2003) Tâches et coopération dans deux dispositifs universitaires de formation à distance, *Apprentissage des Langues et Systèmes d'Information et de Communication* 6(1): 109–25.

Mangenot, F. and E. Nissen (2006) Collective Activity and Tutor Involvement in E-learning Environments for Language Teachers and Learners, *The Calico Journal* 23(3): 601–21.

Mann, S. J. (2004) A Personal Inquiry into an Experience of Adult Learning Online, in P. Goodyear (ed.), *Advances in Research on Networked Learning*, Boston, MA: Kluwer, 205–19.

Mason, R. and A. Kaye (eds) (1989) *Mindweave: Communication, Computers and Distance Education*, Oxford: Pergamon Press (out of print, but available as online publication): http://www.ead.ufms.br/marcelo/ mindware/ mindweave. htm

Mason, R. and A. Kaye (1990) Toward a New Paradigm for Distance Education, in L. Harasim (ed.), *Online Education: Perspectives on a New Environment*, New York: Praeger, 15–38.

Mazur, J. M. (2004) Conversation Analysis for Educational Technologists: Theoretical and Methodological Issues for Researching the Structures, Processes and Meaning of On-line Talk, in D. H. Jonassen (ed.), *Handbook for Research in Educational Communications and Technology*, Mahwah, NJ: Lawrence Erlbaum Associates, 1073–98.

McCambridge, E. (2006) Language Practices of Finnish Sign Language Speakers in Multimodal Environments. Paper presented at 'Integrating CALL into Study Programmes', EUROCALL Conference, Granada, Spain.

McDonough, K. (2006) Action Research and the Professional Development of Graduate Teaching Assistants, *The Modern Language Journal* 90(1): 33–47.

McConnell, D. (2006) *E-Learning Groups and Communities*, Maidenhead: Society for Research into Higher Education and Open University Press.

McHoul, A. W. (1978) The Organisation of Turns at Formal Talk in the Classroom, *Language in Society* 7: 183–213.

McLoughlin, C. and R. Oliver (1999) Pedagogic Roles and Dynamics in Telematics Environments, in M. Selinger and J. Pearson (eds), *Telematics in Education: Trends and Issues*, Amsterdam: Pergamon, 32–50.

Mercer, N., K. Littleton and R. Wegerif (2004) Methods for Studying the Processes of Interaction and Collaborative Activity in Computer-based Educational Activities, *Technology, Pedagogy and Education* 13(2): 195–213.

Meskill, C. (1999) Computers as Tools for Sociocollaborative Language Learning, in K. Cameron (ed.), *Computer Assisted Language Learning: Media, Design and Applications*, Lisse: Swets and Zeitlinger, 141–62.

Meskill, C. (2005) Triadic Scaffolds: Tools for Teaching English Language Learners with Computers, *Language Learning and Technology* 9(1): 46–59.

Meskill, C., J. Mossop, S. DiAngelo and R. K. Pasquale (2002) Expert and Novice Teachers Talking Technology: Precepts, Concepts and Misconcepts, *Language Learning and Technology* 6(3): 46–57.

Mondada, L. (2005) How to Define Corpora for Interactional Analysis. Paper presented at the Contrat Plan Etat Région workshop, Ecole Normale Supérieure des Lettres et Sciences Humaines, Lyons, France.

Mondada, L. (2006) Video Recording as the Reflexive Preservation and Configuration of Phenomenal Features for Analysis, in H. Knoblauch, J. Raab, H.-G. Soeffner and B. Schnettler (eds), *Video Analysis*, Bern: Lang.

Morita, N. (2004) Negotiating Participation and Identity in Second Language Academic Communities, *TESOL Quarterly* 38(4): 573–603.

Mrowa-Hopkins, C. (2000) Une réalisation de l'apprentissage partagé dans un environnement multimédia, *Apprentissage des Langues et Systèmes d'Information et de Communication* 3(2): 207–23.

Murray, L. and T. Hourigan (2008) Blogs for Specific Purposes: Expressivist or Sociocognitivist Approach? *ReCALL* 20(1).

Negretti, R. (1999) Web-based Activities and SLA: A Conversation Analysis Research Approach, *Language Learning and Technology* 3(1): 75–87.

New London Group (1996) A Pedagogy of Multiliteracies: Designing Social Futures, *Harvard Educational Review* 66(1): 60–92.

Nicol, D. J., I. Minty and C. Sinclair (2002) The Social Dimensions of Online Learning, *Innovations in Education and Teaching International* 40(3): 270–80.

Norton, B. (2000) *Identity and Language Learning: Gender, Ethnicity and Educational Change*, Harlow: Longman.

Nunan, D. (1989) *Designing Tasks for the Communicative Classroom*, Cambridge: Cambridge University Press.

O'Dowd, R. (2000) Intercultural Learning via Videoconferencing: A Pilot Exchange Project, *ReCALL* 12(1): 49–63.

O'Dowd, R. (2003) Understanding the Other Side: Intercultural Learning in a Spanish–English Email Exchange, *Language Learning and Technology* 7(2): 118–44.

O'Dowd, R. (2006a) The Use of Videoconferencing and Email as Mediators of Intercultural Student Ethnography, in J. Belz and S. Thorne (eds), *AAUSC 2005 – Internet-mediated Intercultural Foreign Language Education*, Boston, MA: Thomson Heinle and Heinle.

O'Dowd, R. (2006b) *Telecollaboration and the Development of Intercultural Communicative Competence*, Münchner Arbeiten zur Fremdsprachen-Forschung. Band 13. Berlin and München: Langenscheidt.

O'Dowd, R. and M. Ritter (2006) Understanding and Working with 'Failed Communication' in Telecollaborative Exchanges, *The CALICO Journal* 23(3): 623–42.

OECD (2000) Literacy in the Information Age: Final Report of the International Adult Literacy Survey (online publication): http://www1.oecd.org/publications/ e-book/810005le.pdf

Oliver, K. M. (2000) Methods for Developing Constructivist Learning on the Web, *Educational Technology* XL(6): 5–18.

Ong, W. (1982) *Orality and Literacy: The Technologizing of the Word*, London: Routledge.

Open University (2005) *English Grammar in Context, Book 4: Getting down to it: Undertaking Research*, Milton Keynes: The Open University.

Opp-Beckman, L. and C. Kieffer (2004) A Collaborative Model for Online Instruction in the Teaching of Language and Culture, in S. Fotos and C. M. Browne (eds), *New Perspectives on Call for Second Language Classrooms*, Mahwah, NJ: Lawrence Erlbaum Associates, 225–52.

Örnberg Berglund, T. (2005) Multimodality in a Three-dimensional Voice Chat, *Papers in Theoretical Linguistics 92*, Proceedings of the 2nd International Conference on Multimodal Communication, Göteborg. Göteborg: University of Göteborg (online publication): http://blog.humlab.umu.se/therese/files/ 2007/01/multimodality_ornberg.pdf

O'Rourke, B. (2005) Form-focused Interaction in Online Tandem Learning, *The CALICO Journal* 22(3): 433–66.

O'Rourke, B. and K. Schwienhorst (2003) Talking Text: Reflections on Reflection in Computer-mediated Communication, in D. Little, J. Ridley and E. Ushioda (eds), *Learner Autonomy in Foreign Language Teaching: Teacher, Learner, Curriculum, Assessment*, Dublin: Authentik, 47–60.

Ortega, L. (1997) Processes and Outcomes in Networked Classroom Interaction: Defining the Research Agenda for L2 Computer-assisted Classroom Discussion, *Language Learning and Technology* 1(1): 82–93.

Oxford, R. (1997) Cooperative Learning, Collaborative Learning, and Interaction: Three Communicative Strands in the Language Classroom, *The Modern Language Journal* 81(4): 443–56.

Pacagnella, L. (1997) Getting the Seats of Your Pants Dirty: Strategies for Ethnographic Research on Virtual Communities, *Journal of Computer-mediated Communication* 3(1) (online publication): http://www.ascusc.org/jcmc/vol3/ issue1/paccagnella.html

Palfreyman, D. and M. al Khalil (2003) A Funky Language for Teenzz to Use: Representing Gulf Arabic in Instant Messaging, *Journal of Computer-Mediated Communication* 9(1) (online publication): http://jcmc.indiana.edu/vol9/issue1/ palfreyman.html

Panitz, T. (2001) *The Case for Student-Centered Instruction via Collaborative Learning Paradigms* (online resource): http://home.capecod.net/~tpanitz/tedsarticles/ coopbenefits.htm

Paramskis, D. M. (1999) The Shape of Computer-Mediated Communication, in K. Cameron (ed.), *Computer-Assisted Language Learning: Media, Design and Applications*, Lisse: Swets and Zeitlinger, 13–34.

Payne, J. S. and B. M. Ross (2005) Synchronous CMC, Working Memory, and L2 Oral Proficiency Development, *Language Learning and Technology* 9(3): 35–54.

Payne, J. S. and P. J. Whitney (2002) Developing L2 Oral Proficiency through Synchronous CMC: Output, Working Memory, and Interlanguage Development, *The CALICO Journal* 20(1): 7–32.

Pellettieri, J. (2000) Negotiation in Cyberspace: The Role of Chatting in the Development of Grammatical Competence, in M. Warschauer and R. Kern (eds), *Network-based Language Teaching: Concepts and Practice*, Cambridge: Cambridge University Press, 59–86.

Peterson, M. (1997) Language Teaching and Networking, *System* 25(1): 29–37.

Peterson, M. (2001) MOOs and Second Language Acquisition: Towards a Rationale for MOO-based Learning, *Computer-Assisted Language Learning* 14(5): 39–58.

Peterson, M. (2004) MOO Virtual Worlds in CMC-based CALL: Defining an Agenda for Future Research in J.-B. Son (ed.), *Computer-Assisted Language Learning: Concepts, Contexts and Practices*, New York: iUniverse Inc, 39–58.

Piaget, J. (1972) *The Psychology of the Child*, New York: Basic Books.

Pica, T., R. Kanagy and J. Falodun (1993) Choosing and Using Communication Tasks for Second Language Instruction and Research, in G. Crookes and S.M. Gass (eds), *Tasks Language Learning: Integrating Theory and Practice*, Clevedon: Multilingual Matters, 9–34.

Pinkman, K. (2005) Using Blogs in the Foreign Language Classroom: Encouraging Learner Independence, *The JALT CALL Journal* 1(1): 12–24.

Prasolova-Forland, E. and M. Divitni, (2003) Collaborative Virtual Environments for Supporting Learning Communities: An Experience of Use, in Tremaine, M. and C. Simone (eds), *Proceedings of the International ACM SIGGROUP Conference on Supporting Group Work*, Florida, New York: ACM Press, 58–67.

Pujolà, J.-T. (2001) Did CALL Feedback Feed Back? Researching Learners' Use of Feedback, *ReCALL* 13(1): 79–98.

Reffay, C. and T. Chanier (2003) How Social Network Analysis Can Help to Measure Cohesion in Collaborative Distance Learning, in B. Wasson, S. Ludvigsen and U. Hoppe (eds), *Designing for Change in Networked Learning Environments, Proceedings of the International Conference on Computer Support for Collaborative Learning*, Dordrecht: Kluwer: 343–52.

Resnick, L. B. (1991) Shared Cognition: Thinking as Social Practice, in L. B. Resnick, J. M. Levine and S. D. Teasley (eds), *Perspectives on Socially Shared Cognition*, Washington, DC: American Psychological Association, 1–20.

Rice, R.E. (1993) Media Appropriateness: Using Social Presence Theory to Compare Traditional and New Organizational Media, *Human Communication Research* 19(4): 451–84.

Richards, J. C. and T. S. Rodgers (2001) *Approaches and Methods in Language Teaching*, Cambridge: Cambridge University Press.

Riel, M., J. Rhoads and E. Ellis (2006) Culture of Critique: Online Learning Circles and Peer Reviews in Graduate Education, in T. Roberts (ed.), *Self, Peer and Group Assessment in E-Learning*, Hershey, PA: Information Science Publishing.

Roed, J. (2003) Language Learner Behaviour in a Virtual Environment, *Computer-Assisted Language Learning* 16(2/3): 155–72.

Roger, P. (2007) Navigating Research Ethics. http://www.ling.mq.edu.au/research/Navigating%20Research%20Ethics.pdf

Roschelle J. and S. D. Teasley (1995) Construction of Shared Knowledge in Collaborative Problem Solving, in C. O'Malley (ed.), *Computer Supported Collaborative Learning*. Berlin: Springer Verlag, 69–97.

Ros i Solé, C. and M. Truman (2005) Feedback in Distance Language Learning: Current Practices and New Directions, in B. Holmberg, M. A. Shelley and C. J. White (eds), *Distance Education and Languages: Evolution and Change*, Clevedon: Multilingual Matters. 72–91.

Rösler, D. (2004) *E-Learning Fremdsprachen: Eine kritische Einführung*, Tübingen: Stauffenburg.

Rüschoff, B. and M. Ritter (2001) Technology-enhanced Language Learning: Construction of Knowledge and Template-based Learning in the Foreign Language Classroom, *Computer Assisted Language Learning* 14(3/4): 219–32.

Russell, A. L. and L. M. Cohen (1997) The Reflective Colleague in Email Cyberspace: A Means for Improving University Instruction, *Computers and Education* 29(4): 137–45.

Ryan, M.-L. (2003) On Defining Narrative Media, *Image [&] Narrative: Online Magazine of the Visual Narrative* 6 (online publication): http://www.imageandnarrative.be/mediumtheory/marielaureryan.htm

Sacks, H., E. A. Schegloff and G. Jefferson (1974) A Simplest Systematics for the Organisation of Turn-taking for Conversation, *Language* 50(4/1): 696–735.

Salaberry, M. R. (2000) Pedagogical Design of Computer Mediated Communication Tasks: Learning Objectives and Technological Capabilities, *The Modern Language Journal* 84(1): 28–37.

Salmon, G. (2003) *E-moderating: The Key to Teaching and Learning Online*, 2nd edition, London: Routledge.

Santacroce, M. (2004) Analyse du discours et analyse conversationnelle, *Marges Linguistiques* (online publication): http://www.revue-texto.net

Sarangi, S. and C. N. Candlin (eds) (2003) Researching and Reporting the Discourses of Workplace Practice, special issue of *Applied Linguistics* 24(3).

Savignon, S. J. (1997) *Communicative Competence: Theory and Classroom Practice*, 2nd edition, New York: McGraw-Hill.

Savignon, S. J. and W. Roithmeier (2004) Computer-mediated Communication: Texts and Strategies, *The CALICO Journal* 21(2): 265–290.

Sayers, D. (1995) Language Choice and Global Learning Networks: The Pitfall of Lingua Franca Approaches to Classroom Telecomputing, *Educational Policy Analysis* 3(10) (online publication): http://epaa.asu.edu/epaa/v3n10.html

Schmidt, R. W. (1990) The Role of Consciousness in Second Language Learning, *Applied Linguistics* 11: 11–26.

Schneider, J. and S. von der Emde (2006) Conflicts in Cyberspace: From Communication Breakdown to Intercultural Dialogue in Online Collaborations, in J. A Belz and S. L. Thorne (eds), *AAUSC 2005 – Internet-mediated Intercultural Foreign Language Education*, Boston, MA: Thomson Heinle and Heinle, 178–206.

Schön, D. A. (1983) *The Reflective Practitioner: How Professionals Think in Action*, London: Temple Smith.

Schwienhorst, K. (2004) Native-Speaker/Non-Native-Speaker Discourse in the MOO: Topic Negotiation and Initiation in a Synchronous Text-based Environment, *Computer Assisted Language Learning* 17(1): 35–50.

Scollon, R. and S. W. Scollon (1995) *Intercultural Communication*, Oxford: Blackwell.

Scollon, R. and S. W. Scollon (eds) (2003) *Discourses in Place: Language in the Material World*, London: Routledge.

Seedhouse, P. (1998) CA and the Analysis of Foreign Language Interaction: A Reply to Wagner, *Journal of Pragmatics* 30: 85–102.

Seedhouse, P. (2004) *The Interactional Architecture of the Language Classroom: A Conversation Analysis Perspective*, Oxford: Blackwell.

Séror, J. (2005) Computers and Qualitative Data Analysis: Paper, Pens, and Highlighters vs. Screen, Mouse, and Keyboard, *TESOL Quarterly* 39(2): 321–8.

Shelley, M., C. White, U. Baumann and L. Murphy (2006) 'It's a unique role!' Perspectives on Tutor Attributes and Expertise in Distance Language Teaching, *The International Review of Research in Open and Distance Learning* 7(2) (online publication): http://www.irrodl.org/index.php/irrodl/article/view/297/609

Shepard, L. A. (2000) The Role of Assessment in a Learning Culture, *Educational Researcher* 29(7): 4–14.

Sherry, L. (2000) The Nature and Purpose of Online Conversations: A Brief Synthesis of Current Research, *International Journal of Educational Telecommunications* 6(1): 19–52.

Shield, L. (2000) Overcoming Isolation: The Loneliness of the Long Distance Learner. *Wiring the Ivory Tower*, Proceedings of the Millennium Conference of the European Association of Distance Teaching Universities, Paris, Paris: EADTU, 297–302.

Shield, L., M. Hauck and S. Hewer (2001) Talking to Strangers: The Role of the Tutor in Developing Target Language Speaking Skills at a Distance, in A. Kazeroni (ed.), *Proceedings of the UNTELE 2000 Conference, II*, Compiègne: University of Compiègne. 74–84, http://www.utc.fr/~untele/volume2.pdf

Shield, L., M. J. Weininger and L. B. Davies (1999) MOOing in L2: Constructivism and Developing Learner Autonomy for Technology-enhanced Language Learning, *C@lling Japan* 8(3): np (online publication): http://jaltcall.org/cjo/10_99/mooin.htm

Silverman, D. (1997) *Qualitative Research: Theory, Method and Practice*, London: Sage.

Silverman, D. (2001) *Interpreting Qualitative Data: Methods for Analysing Talk, Text and Interaction*. London: Sage.

Simpson, J. (2005) Conversational Floors in Synchronous Text-based CMC Discourse, *Discourse Studies* 7(3): 337–61.

Skehan, P. (1998a) *A Cognitive Approach to Language Learning*, Oxford: Oxford University Press.

Skehan, P. (1998b) Task-based Instruction, *Annual Review of Applied Linguistics* 18: 268–86.

Smith, B. (2005) The Relationship between Negotiated Interaction, Learner Uptake, and Lexical Acquisition in Task-based Computer-assisted Communication, *TESOL Quarterly* 39(1): 33–58.

Snyder, I. (1998) *Page to Screen: Taking literacy into the electronic era*, London: Routledge.

Sotillo, S. (2000) Discourse Functions and Syntactic Complexity in Synchronous and Asynchronous Communication, *Language Learning and Technology* 4(1): 82–119.

Spears, R. and M. Lea (1994) Panacea or Panopticon? The Hidden Power in Computer-mediated Communication, *Communication Research*, 21(4): 427–59.

Stevens, V. (2006) Second Life in Education and Language Learning, *Teaching English as a Second Language-Electronic Journal* 10(3): np (online publication): http://www.tesl-ej.org/ej39/int.html

Stickler, U. and R. Hampel (2007) Designing Online Tutor Training for Language Courses: A Case Study, *Open Learning* 22(1): 75–85.

Stockwell, G. (2003) Effects of Topic Threads on Sustainability of Email Interactions Between Native Speakers and Nonnative Speakers, *ReCALL* 15(1): 37–50.

Svensson, P. (2003) Virtual Worlds as Arenas for Language Learning, in U. Felix (ed.), *Language Learning Online: Towards Best Practice*, Lisse: Swets and Zeitlinger, 123–43.

Svensson, P. (2004) Dispelling the Myth of the Real in Educational Technology. Paper presented at the TeleLearning Research Group Seminar, The Open University.

Swaffar, J., S. Romano, P. Markley and K. Arens (1998) *Language Learning Online: Theory and Practice in the ESL and the L2 Computer Classroom*, Austin, TX: Labyrinth.

Swain, M. (1985) Communicative Competence: Some Roles of Comprehensible Input and Comprehensible Output in its Development, in S. Gass and C. Madden (eds), *Input in Second Language Acquisition*, Rowley, MA: Newbury House, 235–53.

Technorati, Search engine for blogs, http://technorati.com/about

Tella, S. (1999) Firmer Links between Telematics, Multiculturalism and Foreign Language Learning Methodology, in M. Selinger and J. Pearson (eds), *Telematics in Education: Trends and Issues*, Amsterdam: Pergamon, 105–18.

Tella, S. and M. Mononen-Aaltonen (1998) *Developing Dialogic Communication Culture in Media Education: Integrating Dialogism and Technology*, Helsinki: Media Education Publication 7.

Thibault, P. J. (2000) The Multimodal Transcription of a Television Advertisement: Theory and Practice, in A. Baldry (ed.), *Multimodality and Multimediality in the Distance Learning Age*, Campobasso: Palladino, 311–86.

Thorne, S. L. (2003) Artifacts and Cultures-of-use in Intercultural Communication, *Language Learning and Technology* 7(2): 38–67.

Thorne, S.L. (2006) Pedagogical and Praxiological Lessons from Internet-mediated Intercultural Foreign Language Education Research, in J. Belz and S. Thorne (eds), *AAUSC 2005 – Internet-mediated Intercultural Foreign Language Education*, Boston, MA: Thomson Heinle and Heinle, 1–30.

Thorne, S. L. and J. S. Payne (2005) Evolutionary Trajectories, Internet-mediated Expression, and Language Education, *The CALICO Journal* 22(3): 371–97.

Tribble, C. and G. Jones (1997) *Concordances in the Classroom: A Resource Book for Teachers*, Houston, TX: Athelstan.

Truscott, S. and J. Morley (2001) Cross-cultural Learning through Computer-mediated Communication, *Language Learning Journal* (24): 17–23.

Tu, C. H. and M. McIsaac (2002) The Relationship of Social Presence and Interaction in Online Classes, *The American Journal of Distance Education.* 16(3): 131–50.

Tudini, V. (2003) Using Native Speakers in Chat, *Language Learning and Technology* 7(3): 141–59.

van Lier, L. (1996) *Interaction in the Language Curriculum: Awareness, Autonomy and Authenticity*, Harlow: Longman.

van Lier, L. (2000) From Input to Affordance, in J. Lantolf (ed.), *Sociocultural Theory and Second Language Learning*, Oxford: Oxford University Press, 245–59.

van Lier, L. (2002) An Ecological-Semiotic Perspective on Language and Linguistics, in C. Kramsch (ed.), *Language Acquisition and Language Socialization: Ecological Perspectives*, London and New York: Continuum Books, 140–64.

Varonis, E. M. and S. Gass (1985) Non-native/Non-native Conversations: A Model for Negotiation of Meaning, *Applied Linguistics* 6(1): 71–90.

Vetter, A. (2003) Instructor's Reflective Diary, personal communication.

Vetter, A. (2004) Les Spécificités du tutorat à distance à l'Open University: enseigner les langues avec Lyceum, *Apprentissage des Langues et Systèmes d'Information et de Communication* 7: 107–29.

Vogiazou Y., M. Dzobor, J. Komzak and M. Eisenstadt (2003) BuddySpace: Large-scale Presence for Communities at Work and Play. Paper presented at the International Conference on Communities and Technologies, Amsterdam (online publication): http://kmi.open.ac.uk/publications/pdf/kmi-03-14.pdf

Vogiazou, Y., M. Eisenstadt, M. Dzbor and J. Komzak (2005) From Buddyspace to CitiTag: Large-scale Symbolic Presence for Community Building and Spontaneous Play, *Proceedings of the 2005 ACM Symposium on Applied Computing*, Santa Fe, New Mexico (online publication): http://portal.acm.org/ citation.cfm?id=1067040

von der Emde, S., J. Schneider and M. Kötter (2001) Technically Speaking: Transforming Language Learning through Virtual Learning Environments (MOOs), *The Modern Language Journal* 85(2): 210–25.

Vygotsky, L. S. (1978) *Mind in Society: The Development of Higher Psychological Processes*, Cambridge, MA: Harvard University Press.

Wagner, J. (1996) Language Acquisition through Foreign Language Interaction: A Critical Review of Studies in Second Language Acquisition, *Journal of Pragmatics* 26: 215–35.

Walther, J. B. (1996) Computer-mediated Communication: Impersonal, Inter-personal and Hyperpersonal Interaction, *Communication Research* 23(1): 3– 43.

Ward, J. M. (2004) Blog Assisted Language Learning (BALL): Push Button Publishing for the Pupils. *TEFL Web Journal* 3(1) (online publication): http://www.teflweb-j.org/v3n1/blog_ward.pdf

Ware, P. D. (2003) From Involvement to Engagement in Online Communication: Promoting Intercultural Competence in Foreign Language Education, unpublished doctoral thesis, University of Arizona.

Warner, C. (2004) It's Just a Game, Right? Types of Play in Foreign Language CMC, *Language Learning and Technology* 8(2): 69–87.

Warschauer, M. (1995) *Virtual Connections: Online Activities and Projects for Networking Language Learners*, Honolulu: Second Language Teaching and Curriculum Centre, University of Hawaii.

Warschauer, M. (1997) Computer-mediated Collaborative Learning: Theory and Practice, *The Modern Language Journal* 81(4): 470–81.

Warschauer, M. (1998) Online Learning in Sociocultural Contexts, *Anthropology and Education Quarterly* 29(1): 68–88.

Warschauer, M. (1999a) *Electronic Literacies: Language, Culture, and Power in Online Education*, Mahwah, NJ: Lawrence Erlbaum Associates.

Warschauer, M. (1999b) CALL vs. Electronic Literacy: Reconceiving Technology in the Language Classroom. Paper presented at the Centre for Information on

Language Teaching and Research Annual Research Forum, Cambridge, 1998 (online publication): http://www.cilt.org.uk/research/papers/resfor2/warsum1. htm

Warschauer, M. (2000) On-line Learning in Second Language Classrooms: An Ethnographic Study, in M. Warschauer and R. Kern (eds), *Network-based Language Teaching: Concepts and Practice*, Cambridge: Cambridge University Press, 41–58.

Warschauer, M, and D. Healey (1998) Computers and Language Learning: An Overview, *Language Teaching Research* 31(2): 57–71.

Warschauer, M. and R. Kern (eds) (2000) *Network-based Language Teaching: Concepts and Practice*, Cambridge: Cambridge University Press.

Warschauer, M. and S. Lepeintre (1997) Freire's Dream or Foucault's Nightmare? Teacher–Student Relations on an International Computer Network, in R. Debski, J. Gassin and M. Smith (eds), *Language Learning through Social Computing*, Melbourne: University of Melbourne, 67–89.

Warschauer, M., L. Turbee and B. Roberts (1996) Computer Learning Networks and Student Empowerment, *System* 24(1): 1–14.

Weasenforth, D., S. Biesenbach-Lucas and C. Meloni (2002) Realizing Constructivist Objectives through Collaborative Technologies: Threaded Discussions, *Language Learning and Technology*, 6(3): 58–86.

Webopedia (online resource): http://www.webopedia.com/TERM/b/blog.html.

Weininger, M. J. and L. Shield (2003) Promoting Oral Production in a Written Channel: An Investigation of Learner Language in MOO, *Computer Assisted Language Learning* 16(4): 329–49.

Weir, C. (2005) *Language Testing and Validation: An Evidence-based Approach*, Basingstoke: Palgrave Macmillan.

Wenger, E. (2005) Communities of Practice: A Brief Introduction (online publication): http://www.ewenger.com/theory/communities_of_practice_intro.htm

Wertsch, J. V. (1991a) *Voices of the Mind: A Sociocultural Approach to Mediated Action*, London: Harvester Wheatsheaf.

Wertsch, J. V. (1991b) A Sociocultural Approach to Socially Shared Cognition, in L. B. Resnick, J. M. Levine and S. D. Teasley (eds), *Perspectives on Socially Shared Cognition*, Washington, DC: American Psychological Association, 85–100.

Wertsch, J. V. (2002) Computer Mediation, PBL, and Dialogicality, special issue of *Distance Education* 23(1): 105–8.

White, C. (2003) *Language Learning in Distance Education*, Cambridge: Cambridge University Press.

Whitelock, D. (2006) Electronic Assessment: Marking, Monitoring and Mediating Learning, *International Journal of Learning Technology* 2(2/3): 264–76.

Wiki, *Webopedia* (online resource): http://www.webopedia.com/ TERM/ w/wiki. html

Wikipedia: The Free Encyclopedia (online resource): http://en.wikipedia.org/wiki/Main_Page

Williams, L. F. (2003) The Nature and Complexities of Chat Discourse: A Qualitative Case Study of Multi-level Learners of French in an Electronic Environment, unpublished doctoral thesis, Pennsylvania State University.

Wolfe, C. R. (ed.) (2001) *Learning and Teaching on the World Wide Web*, Educational Psychology Series, San Diego: Academic Press.

Wooffitt, R. (2005) *Conversation Analysis and Discourse Analysis: A Comparative and Critical Introduction*, London: Sage.

Wu, W. S. (2005) Using Blogs in an EFL Writing Class. Paper presented at the 2005 Conference and Workshop on TEFL and Applied Linguistics (online publication): http://www.chu.edu.tw/~wswu/publications/papers/conferences/05.pdf

Yule, G. (1997) *Referential Communication Tasks*, Mahwah, NJ: Lawrence Erlbaum Associates.

Zähner, C., A. Fauverge and J. Wong (2000) Task-based Language Learning via Audiovisual Networks? In M. Warschauer and R. Kern (eds), *Network-Based Language Teaching: Concepts and Practice*. Cambridge: Cambridge University Press, 186–203.

Zemsky, R. and W. Massey (2004) *Thwarted Innovation: What Happened to e-Learning and Why*, The Learning Alliance at the University of Pennsylvania with Thomson Corporation.

Zhang, R. (2004) Using the Principles of Exploratory Practice to Guide Group Work in an Extensive Reading Class in China, *Language Teaching Research* 8: 331–45.

Zhao, Y. (2003) Recent Developments in Technology and Language Learning: A Literature Review and Meta-analysis, *The CALICO Journal* 21(1): 7–27.

Zuengler, J. and E. R. Miller (2006) Cognitive and Sociocultural Perspectives: Two Parallel SLA Worlds? *TESOL Quarterly* 40(1): 35–58.

Index